Pioneer Settlers in Nineteenth-Century Brazil

English, Irish and Irish-American Pioneer Settlers in Nineteenth-Century Brazil

Oliver Marshall

Centre for Brazilian Studies
University of Oxford

Oliver Marshall *is a Research Associate at the Centre for Brazilian Studies, Oxford. His publications include* The English-Language Press in Latin America *(London: Institute of Latin American Studies, University of London, 1996),* Brazil in British and Irish Archives *(Centre for Brazilian Studies, University of Oxford, 2002) and (as editor)* English-Speaking Communities in Latin America *(London: Macmillan, 2000).*

© Oliver Marshall, 2005

Published by
Centre for Brazilian Studies
University of Oxford
92 Woodstock Road
Oxford OX2 7ND

ISBN 0-9544070-4-0

Cover illustration: "The Grange" – the Tamplin family's property alongside the Turvo River, Assunguy, Paraná, in the early 1870s, by Caroline Tamplin. Courtesy of the Tigar family, Vancouver, Canada.
Cover design by Andrew Chapman, Chapman Design.
Maps by David Sansom.
Typeset by Meg Palmer, Third Column.
Printed by Lightning Source.

Table of Contents

List of maps and illustrations   vii

List of tables   viii

Acknowledgements   ix

PART I

Prologue   3

Introduction   7

PART II

Chapter 1   Agricultural Colonization in Brazil, 1808–67   13

PART III

Chapter 2   Brazil and the Irish Diaspora   35

Chapter 3   A "New Ireland" in Brazil   63

PART IV

Chapter 4   Agricultural Labourers in Mid-Victorian England   91

Chapter 5   "The Workhouse – or Brazil": The Recruitment of English Emigrants   103

Chapter 6   From England to the Brazilian Colonies   117

Chapter 7   Pioneering in South Brazil   137

Chapter 8   The Collapse   171

PART V

Conclusion _____ 191

Epilogue _____ 207

APPENDICES

Appendix 1 A petition to Pope Pius IX_____ 219

Appendix 2 1898: A final appeal for assistance _____ 225

Appendix 3 British immigrants in Príncipe Dom Pedro,
               Cananéia and Assunguy_____ 227

Notes_____ 269

Picture Credits _____ 303

Bibliography _____ 305

Index_____ 317

Maps and Illustrations

Map of England	xii
Map of southern Brazil	xiii
A Black Country townscape	55
Barziller Cottle in 1867	71
Field hands taking a refreshment break	95
Labourers' Union Chronicle, 18 January 1873	104
Agricultural labourers meeting in Wellesbourne, Warwickshire, 1872	107
The emigrant's last sight of home	115
Fazenda Montebello, Espírito Santo	129
A panic upon a bridge	154
The first stage of clearing the forest for settlement	160
'The Grange': the Tamplin family's farm	163
The Turvo Valley schoolhouse	168
A wedding gathering, Cêrro Azul	209
Three generations of the Chamberlain family	210
Luiza da Conceição Blane and Ernesto Fitz	211
João and David Davies	215

Tables

Table 1:
Immigrant entries to the port of Rio de Janeiro, 1864–73 _____ 21

Table 2:
Population of Wednesbury and the Black Country, 1841–71 ____ 45

Table 3:
Nationality of immigrants arriving at Côlonia Príncipe
Dom Pedro, Jan–Dec 1867 _____ 73

Table 4:
Nationality of immigrants arriving at Côlonia Príncipe
Dom Pedro, Feb–Oct 1868 _____ 73

Table 5:
Nationality of immigrants arriving at Côlonia Príncipe
Dom Pedro, Jan–April 1869 _____ 74

Acknowledgements

This book has been with me for rather a long time. In acknowledging the help and support that I have received, I hope that I have not lost track of too many people and institutions who have provided assistance as I fitfully worked on this project.

Crucially, thanks are due to countless librarians, archivists and curators in Brazil, England and the United States. Apart from the staffs of the archival repositories listed in the bibliography on page 305, I am grateful to those at the Biblioteca Nacional (Rio de Janeiro), University of London Library (Senate House), Canning House Library (the Hispanic and Luso-Brazilian Council, London), Smethwick Library and the Warwickshire County Record Office. Although offering a sanitized version of the past, the Black Country Living History Museum in Dudley was inspiring to visit, not least because my daughter, Anneliese Doyle, so enthusiastically shared my interest in the reconstruction of a nineteenth-century south Staffordshire village.

One of the most enjoyable aspects of researching this book has been corresponding with, and meeting, descendants of some of the immigrants who feature in this study. I am especially grateful to Mauro Fortes Carneiro (of Curitiba), whose great-grandfather, Ernest Bond, emigrated to Brazil in 1872 at the age of seven. In addition to helping me to piece together the story of the Bond family, Mauro provided valuable genealogical information on other English families who settled in Paraná and answered innumerable questions concerning his city and state.

Other descendants of immigrants generously shared with me family documents: thanks are due to Helen Jean Gent (Pittsburgh) for copies of letters sent by her great-grandmother to relatives in England; Percy Tamplin (Curitiba) for allowing me to examine his great-grandmother's diaries; and Arlene Tigar Maclean (Vancouver) for a typescript of her grandfather's memoirs and reproductions of her great-great-grandmother's watercolours, one of which is reproduced on the cover and within the pages of this book.

For providing details of family history, thank are also owed to Margarida Chamberlain, Auta Desplanches, and Hilda Maria Pugsley Osti (all of Curitiba), Carlito Chamberlain (Cêrro Azul), David and João Davies (Cananéia), Antonio Davies (Iguape), Ivan Davies Martins (São Paulo), Carlos Eduardo Young (Rio de Janeiro) and Dave Alsop, Rob Ettridge and John Watts (England). In addition, genealogists David Aspey and Anthony McCan provided information on the Irish background of William Scully.

I have benefited enormously from contacts with other scholars. In particular I am grateful to Leslie Bethell for encouraging me to complete this project and for the institutional support of the Centre for Brazilian Studies, Oxford. Members of the Irish Diaspora Studies scholarly network (irishdiaspora.net) – and in particular its moderator, Patrick O'Sullivan – never failed to respond to my queries, enabling me to attempt to understand how 'my' Irish migrants relate to the wider Irish Diaspora. I am also grateful to Miguel Alexandre de Araujo Neto, Rosana Barbosa, Roderick Barman, Aloisius Carlos Lauth, Brian McGinn and Edmundo Murray for entering into discussions and for offering advice regarding aspects of nineteenth-century Brazil, international migration and the Irish Diaspora. From early days, Eduardo Silva has also been especially encouraging and has pointed me towards many crucial texts. Eduardo, Graça and Isaura have on many occasions generously opened their home in Humaitá, providing all kinds of advice on Rio, the most wonderful Bahian food and warm company.

The book would not have been possible without the support of my family. My parents, Peter and Liselotte Marshall, have been interested in the book's progress, with each providing very practical assistance. My mother translated into English articles from the newspaper *Die Kolonie Zeitung* and sections of the German-language books that are listed in the bibliography. My father, a historian of North America, took a great interest in my research, suggested many sources and carefully read, commented on and edited the entire text. I am enormously grateful to them both.

Finally, my wife, Margaret Doyle, deserves more than my gratitude. She has endured hearing the stories of 'my' emigrants and witnessing my tortuous writing process for far too long.

Oliver Marshall
London – February 2005

Map 1: England

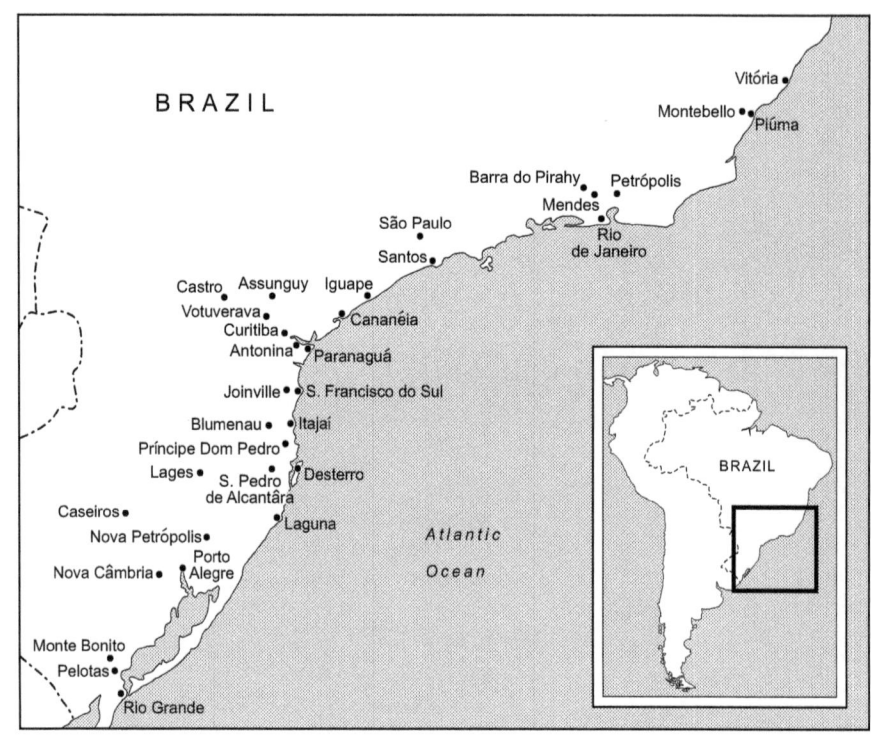

Map 2: Southern Brazil

Part I

Prologue

IN THOMAS HARDY'S novel *Tess of the D'Urbervilles,* Angel Clare emigrates to Brazil. For Angel's father, the answer to the question faced in the nineteenth century by parents of young British 'gentlemen' – 'What shall we do with the boys?' – was simple: Angel and his brothers would follow in his footsteps by attending Cambridge University before going on to take Holy Orders. Rebelling against such expectations, Angel chooses to enter farming, with a view to settling in the English Midlands, a British colony or Texas. To prepare himself for his new life as an agriculturalist, the young man spends periods of pupilage, living and working alongside farm hands.

While on a Wessex dairy farm, Angel falls in love with a milkmaid, Tess Durbeyfield, daughter of a poor and drunken peddler and carter who is descended from the D'Urbervilles, a once wealthy and noble family. On their wedding night, Tess tells Angel how, at the age of fifteen, she had been seduced and bore a child who died in infancy. Angel is mortified by the revelation (" 'Here was I thinking you a new-sprung child of nature; there were you, the belated seedling of an effete aristocracy!' "[1]) and abandons Tess to re-evaluate his outlook on life. For three weeks, Angel travels the Dorset countryside in a trance-like state, and at some point notices a placard advertising the advantages of Brazil for emigrating agriculturists, land being available on apparently excellent terms. The idea of Brazil at once attracts Angel: Tess could eventually join him, "and perhaps in that country of contrasting scenes and notions and habits the conventions would not be so operative which made life with her seem impracticable [in England]".[2] Despite the shocked reaction of his pious Anglican parents – " 'Brazil! Why they are all Roman Catholics there surely!' "[3] – and the discouraging reports filtering home from farm labourers who had recently emigrated there,

Angel flees alone to South America.

After lying ill with fever in "the clay lands near Curitiba", and having been "drenched with thunder-storms and persecuted by other hardships, in common with all the English farmers and farm-labourers who, just at this time, were deluded into going thither by the promises of the Brazilian Government, and by the baseless assumption that those frames which, ploughing and sowing on English uplands, had resisted all the weathers to whose moods they had been born, could resist equally well all the weathers by which they were surprised on Brazilian plains", Angel resolves to return to Tess.[4] His experiences of "this strange land" had proved hopeless and he abandons his plan to farm there. Angel sets off on foot towards the coast to seek a passage home, and on the trail he encounters fellow English immigrants:

> The crowds of agricultural labourers who had come out to the country in his wake, dazzled by representations of easy independence, had suffered, died, and wasted away. He would see mothers from English farms trudging along with their infants in their arms, when the child would be stricken with fever and would die; the mother would pause to dig a hole in the loose earth with her bare hands, would bury the babe therein with the same natural grave-tools, shed one tear and again trudge on.[5]

The horrors Angel experiences during his few months in Brazil age him by a dozen years and provoke him to re-evaluate his outlook on life, not least his previous harsh judgement of Tess. On arriving back in England, Angel is in a pitiful state:

> [S]o reduced was that figure from its former contours by worry and the bad season that Clare had experienced, in the climate to which he had so rashly hurried in his first aversion to the mockery of events at home. You could see the skeleton behind the man, and almost the ghost behind the skeleton [. . .] His sunken eye-pits were of morbid hue, and the light in his eyes had waned. The angular hollows and lines of his aged ancestors had succeeded to their reign in his face twenty years before their time.[6]

In the book *Ingleses,* a collection of essays published in 1942, the Brazilian sociologist Gilberto Freyre was scathing of Hardy's presentation of Brazil. In contrast, Freyre totally accepted the widely-held Brazilian view that in the nineteenth century Paraná (the Brazilian province where Hardy placed Angel Clare) offered "idyllically favourable" conditions to "even the most delicate of immigrants", and that the few obstacles in the way of settlement were no greater than those encountered by pioneers in "the finest areas of North America". Freyre argued that Hardy had created a Paraná far removed from any geographical or historical reality, at best confusing equatorial conditions in the Amazonian province of Pará with those of the temperate south of the country.[7]

While Thomas Hardy was, without question, well acquainted with the often-desperate plight of labourers in the English West Country, do his accounts of the conditions endured by English settlers in Brazil owe more to fantasy than reality? Are Hardy's descriptions of Brazil simply literary devices, metaphors for insanity into which he let slip basic errors in geography, as well as ignorance and prejudice towards a "Papistical land"? It is certainly a fact that many hundreds of British immigrants were attracted to Brazilian agricultural colonization schemes in the 1860s and 1870s, years that form the background of so much of Hardy's writing, but how did their actual experiences compare to the conditions encountered by the fictional Angel Clare? Examining the backgrounds of the schemes and of people attracted to them, such questions will be amongst those considered in this study of English, Irish and Irish-American pioneer settlers in nineteenth-century Brazil.

Introduction

BEFORE AN IMMIGRANT-BASED community can assert its identity, be influential or assimilate, it has to be established. Before it can be established, it has to be imagined. Brazil is not usually associated with either English or Irish immigrants, but in the late 1860s and early 1870s great efforts were made in England and elsewhere to stimulate interest in the country. An idealized image of Brazil was created to help persuade dissatisfied Irish and English to pack up and join agricultural settlement schemes in a country that they had known nothing about, the recruits carrying unrealistic expectations of what lay before them.

Tracing the establishment – or failure – of an immigrant community requires not merely an understanding of the conditions that they experienced in the host country, but also an understanding of how the choice of destination was arrived at and what experiences, expectations and ambitions migrants carried with them. More than any other potential destination at this time, Brazil – and South America in general – was an 'exotic' one for British emigrants; little was known about the reality of life there. Instead the prevalent images of Brazil, among those English and Irish who were able to conjure up even a limited mental picture of the country, were of a land of mystery or a lush paradise, a place on which, according to a guide book aimed at immigrants and merchants, "nature, in the collocation and accumulation of its mineral and agricultural wealth, seems to have smiled benignantly [sic], and to have lavished with a munificent hand her choicest treasures".[1]

South America's very otherness meant that most of the continent's immigration schemes that involved Britons were based on ignorance. Some of these were no more than speculative business ventures, and most exploited the often-desperate circumstances that would-be emigrants were living under in England.[2] Other destinations – such as North

America, Australia and New Zealand – were somehow more imaginable to the British; everything from the climate to the customs, legal system and language seemed much closer to their own. Literacy in nineteenth-century Britain was far from universal, but newspapers – never short of articles arguing for or against particular destinations – were widely available, as were specialist promotional magazines, pamphlets and guide books, their contents spread by those able to read them. But for most people, the most credible accounts of far-off lands were found in the letters that flowed continuously between friends and relatives abroad, news that was passed to friends and neighbours at home but was often also forwarded to local newspapers for publication.[3] Migration, moreover, was not just a one-way traffic, English return migration rates being consistently amongst Europe's highest.[4] As such, it is reasonable to assume that those returning related their experiences, positive and negative, to folk back home, providing vivid and personalised images of their emigration experiences. Especially grounded in reality were the expectations of migrants destined for North America, based, as they were, on varied and trusted (though ultimately not always reliable) sources. This meant that the emigrants were better prepared for what lay ahead, while the weight of numbers allowed for the creation of large personal networks that enabled newcomers to better defend themselves if things went wrong in the main immigrant-receiving countries.

In the Brazilian settlement schemes of the 1860s and 1870s, several factors combined to create impossible expectations on the part of the emigrants. The Brazilian government and its overseas diplomatic missions, recruiters (English and Brazilian agents representing – or sometimes merely purporting to represent – Brazil), and even elements of the Catholic church and labourers' unions all had interests in promoting the schemes, albeit for widely varying reasons. For the most part, the people they set out to target were simple labourers: people who had neither the capital nor the experience to be independent farmers whether at home or in most other available immigrant-receiving destinations. The living and working conditions that these labourers endured in England were always difficult (frequently they were horrendous), seemingly a mirror image of the picture presented of Brazil. Yet it was not only labourers who were taken in by the claims made about the schemes; many elements of the national and local press were also deceived, albeit almost wilfully in a few instances. They readily and enthusiastically took up the call to promote Brazil as a land of unparalleled potential prosperity for immigrants – without much evidence to support such a view. Letters, articles, pamphlets and books by earlier British immigrants and travellers, extolling the virtues of the Brazilian paradise, were published with their accuracy – or the motives behind them – only rarely questioned. Some of these writings were produced by long-time British residents of Rio de Janeiro keen to support the Brazilian government's quest to encourage northern European immigrants. Others were penned by some of the first of the wave of British labourers attracted to Brazil, people who might well have been feeling intensely physically and culturally isolated – some had not even realized they were going to a country where English was not spoken – and who were desperate to have compatriots boost their numbers. It was not in their interest to paint a realistic portrait of the hardships and disappointments of life in Brazil.

Despite the rapid and hardly surprising (at least in retrospect) collapse of English, Irish and Irish-American settlement attempts in Brazil, these episodes are worth recalling for two primary reasons. From the Brazilian perspective, these settlement attempts address issues of immigration policy and the execution of the colonization process alongside related concerns of race, modernization and the projection of Brazil's image abroad. In a broader sense, the backgrounds in England – and also the United States – of those English and Irish people who were lured into the Brazilian wilderness serve as a stark reminder that the migration process often involves more than a simple one-way passage. For many individuals, it essentially involved moving from one place to a second, but for a substantial proportion the process involved further upheavals, leading to return or onward migration either internally or to third countries.

Return migration is an issue that has been increasingly addressed by historians with, for example, examinations of transatlantic seasonal labour transfers and 'successful' and 'failed' returnees.[5] However re-emigration – termed "experimental mobility" by Philip Taylor, historian of United States and international migration – is rarely a focus of study, or even discussed, although the topic has long been on the research agenda.[6] Many of those involved in such transfers to third countries were motivated by what would prove to be genuinely better conditions such as improved economic circumstances, a greater cultural affinity or opportunities of being reunited with friends or relatives. But onward migration could also be a result of desperation, involving people who would flee an environment with few apparent prospects and be ready to accept offers of land or employment elsewhere despite a lack of pre-existing social ties to call upon for advice or, if help was needed, support.[7]

How many British emigrants – whether 'agriculturalists' or otherwise – were recruited overall is unrecorded. Many were dispatched in small groups aboard ships that, with rare exceptions, carried too few passengers to bring them under the terms of the 1855 Passenger Act whereby captains were obliged to provide port authorities with passenger lists which, in any case, have not survived. But within the space of five years, certainly several thousand English and Irish labourers (along with a small number of men and families of higher social standing) were dispatched to Brazil, with 1868 and 1872–3 being the years of most intensive recruitment.[8] An examination of these efforts and the factors that combined to make them at least a short-term success for the recruiters, though frequently dismal failures for those recruited, is enlightening. While there certainly were successes that did much to mould the social and economic characters of areas of Brazil, the disastrous settlement schemes were held out for decades as a reminder to people not to believe all that was said by emigration agents. For Brazil, an enormous expenditure on state agricultural colonies that resulted in only very patchy success, attracting just a few thousand immigrants a year, caused immense harm to the reputation of the country. This made the Brazilian government extremely wary of ever again being so directly involved in the recruitment and settlement of immigrants.

Part II

The people of these islands are more movable than other nations, and large numbers of them are always abroad, sometimes on distant voyages, sometimes on the Alps, sometimes in the deserts of Africa, or in the strangest of places....

Census of England and Wales, General Report, (1861), vol. 3, p. 4.

As you are aware, there are so many British colonists scattered all over the face of the globe, so many of whom unfortunately have reason to be dissatisfied with their lot, that it would be impossible to take their cases, except under very exceptional circumstances, with favourable consideration.

British consul in Rio de Janeiro
to John Mitchell, British *colono* in Cananéia
21 December 1880
[NA/PRO FO128/116]

Chapter 1
Agricultural Colonization in Brazil, 1808–67

UNTIL THE NINETEENTH century, Latin America received an insignificant number of immigrants from Britain and Ireland. Although not entirely off-limits to non-Spanish or non-Portuguese subjects, a combination of factors, including restrictions on foreign trade and investment, suspicion and legal barriers, meant that many of those British and Irish adventurers who ventured there had previously been living in peninsular Spain or Portugal, the Canary Islands, Madeira or the Azores. Some of these were independent traders, while others were civil servants or military officers employed by the Spanish or Portuguese governments.

With the political changes leading to independence that swept Latin America from the beginning of the nineteenth century, Britain rapidly emerged as the region's pre-eminent economic power. The transfer under British naval escort of the Portuguese royal family and Portugal's entire state apparatus to Rio de Janeiro in 1807–8, in advance of Napoleon's invading army, immediately resulted in a change in the relationship between Brazil and the rest of the world. Brazilian ports were now opened to direct international trade – in practice, due to the war in Europe, only with England – having previously had all trade channelled through Portugal. Britain's dominant economic position was formalized by the new government in Rio who rewarded it with preferential trading rights, limiting tariffs that could be levied on British imports. By the time Brazilian Independence was recognized internationally in 1824, Brazil had become Britain's third largest export market.[1]

In Brazil (as in most other parts of Latin America) British people were always few in number but from the early years of the nineteenth century they formed a highly visible and influential presence. Amongst the first to be established were merchants who dominated the import and export trades, with Recife (Pernambuco), Salvador (Bahia) and, in particular, Rio de Janeiro being the most important centres of business activity. As the century progressed Britain's economic interests expanded to include mining, shipping services, port facilities, railways, utility companies and banking. A consequence of British dominance in these expanding sectors was that Britain became synonymous in the minds of many Brazilians with modernization – and English-speakers came to be considered as agents of progressive change.[2]

The British and Irish who settled in Brazil were diverse in background. Lured by bounties and the hope of prize money, men were recruited for the highest and lowest positions in Brazil's army and navy; merchants established themselves in ports from Pará in the north to Rio Grande in the far south; and armies of managers and engineers of all levels were occupied in the mines, sugar mills, towns and cities, overseeing Britain's general economic penetration of the country. The British communities that came into being were closely-knit, with every effort made to remain aloof from Brazilian society other than for the purposes of work. Lowly clerks hoping to scale the ranks of their companies and skilled railway, mine and other contract workers certainly existed in large numbers, but they usually lived apart from their wealthy compatriots and were regarded as transients and therefore most definitely not to be considered as part of the mainstream British communities. The image of prosperity and attitudes of social superiority and detachment were encouraged, being seen to advance the prestige of the community, whereas a too obvious presence of lower-class Britons could prove an embarrassment and inconvenience.[3]

British agricultural immigrants in Brazil were, during the nineteenth century, usually few in number and certainly not the conspicuous and economically important presence that they were on the Argentine pampas or, further south still, in Patagonia. A handful of men from Britain and Ireland established sugar plantations in Pernambuco and Bahia, coffee plantations in the province of Rio de Janeiro or were involved in raising cattle in the grasslands of Rio Grande do Sul. But in general, even with the necessary levels of capital to invest, prospects in such areas were limited as a result of local oligarchies' suspicions of outsiders and British governmental restrictions on the ownership of slaves by British subjects.

But throughout the nineteenth, and even well into the twentieth century, Latin American governments encouraged immigrant-based agricultural settlements. Although virtually every country in the region experienced immigrant-based land colonization schemes at some point since independence, it was Brazil that was the most persistent in efforts to attract pioneers. These schemes, both private and state sponsored, had various driving forces. Some were first and foremost geopolitical tools, asserting the authority of the state and boosting population in border areas and under-populated parts of the interior beyond effective government control; others were essentially speculative business ventures by private land development companies. Many of these schemes were also perceived as having the higher purpose of introducing rational or scientific methods of agricultural production based on family farms of the sort that northern Europeans were considered to have perfected, a sentiment that Carlos Perret Gentil, the Swiss consul in Rio de Janeiro, expressed in 1851:

> Immense are the advantages deriving from the introduction of a white population. With it superior labor and progressive intelligence are attracted; settlement nuclei are created and developed; industry is born; agriculture grows; lands increase in value; and there is no need to employ great amounts of capital, as with the slaves.[4]

During the nineteenth century as a whole, Brazil ranked amongst the top three immigrant-receiving countries in the Americas, with only the United States and Argentina attracting more settlers. But it was only with the progressive decline of slavery in the late 1870s and 1880s that large-scale immigration really got underway. Not until 1888 and the final abolition of slavery in Brazil did the country become a major immigrant-receiving destination, able to take advantage of the emergence of southern Europe as a major source of emigrants. That said, numerous efforts were made earlier in the century to attract immigrants. Many quickly ended in failure, but some had profound local and regional impacts that even today remain apparent.

There had long been a steady flow of people between mainland Portugal, Madeira and the Azores, to and from Brazil. Not until the transfer of the Portuguese government to Rio de Janeiro in 1807 was the vast South American territory gradually opened to non-Portuguese immigrants. But until the 1880s Portuguese immigrants continued to represent the bulk of arrivals, although they generally remained in the cities, especially Rio. In 1810 and in 1812, several hundred Chinese indentured labourers were shipped to Rio with the aim of establishing tea plantations. Experimental efforts just outside the city limits rapidly collapsed, exacerbated by labourers falling victim to fatal illnesses, with the survivors being turned out into the streets of the Brazilian capital or somehow finding their way into the interior to labour in the mines of Minas Gerais.[5] But it was only in 1818 that non-Portuguese free immigration seriously commenced, with the establishment of the German farming settlement of Leopoldina, in the province of Bahia, followed shortly after by the more enduring (although often troubled) Swiss colony of Nova Friburgo, situated in the interior of the province of Rio de Janeiro.[6] From this time, systematic efforts were made to attract immigrants to settle as independent smallholders in under-populated, economically marginal but often strategically significant, parts of the country. Initially, the overwhelming bulk of the immigrants who participated in the agricultural colonization schemes were, as in Leopoldina and Nova Friburgo, either German or Swiss. Most immigrant arrivals at about this time were provided with land grants in Brazil's sub-tropical and temperate south, although several enduring agricultural colonies (as well as some that very quickly disintegrated) were located well within the tropics, most notably in the province of Espírito Santo, but also, even closer still to the equator, in the provinces of Bahia and Pará.[7]

There was also considerable debate regarding the possible introduction of immigrants to work as sharecroppers on the expanding coffee-growing plantations in the province of São Paolo as the owners came to realize that they could not rely on importing African slaves to meet their labour requirements. Discussion relating to potential immigrants often revolved around issues of race, falling back on determinist arguments to help identify which races, ethnic groups or nationalities would prove the most positive influences in moulding Brazilian society.[8] The cultural historian Víanna Moog has explained the Brazilian elite's political and intellectual pseudoscientific thesis advocating the direction that should be taken in moulding the country's racial makeup:

> Brazil's backwardness in comparison with the United States was inherited in her people's Portuguese descent, and also, and principally, to the debasing mixture of the Portuguese with the two other races, Indian and Negro, that entered into the founding of the nation. In this way everything was explained, everything perfectly solved. Since the white race alone could create progress, the sole hope for the inferior races was to continue intermarrying with the superior whites.[9]

1. Agricultural Colonization in Brazil, 1808–67

In the early 1850s, Luiz Peixoto de Lacerda Werneck, a wealthy and politically influential plantation-owner in Rio de Janeiro's rich coffee-producing Paraíba Valley, articulated the Brazilian élite's generally held preference for German immigrants.[10] Werneck offered stark warnings against looking to Asia – in his opinion the homeland of "barbarous nations [inhabited by] races decrepit in spirit and deformed in body" – as a labour pool to meet the anticipated impending explosion in demand resulting from the cessation of the Atlantic slave trade. Chinese were beginning to be shipped to Latin America (in particular to Cuba and Peru) in considerable numbers as indentured labourers, and East Indians were being transported to the British West Indies (in particular to Trinidad and British Guiana), but Werneck argued strongly against any such migrants being introduced to Brazil. To Werneck, Chinese and other Asians were considered as being members of a "stationary, a doubtful civilization" and people "inert from progress". "We trust in God," pleaded Werneck, "that they will not be turned loose on the Brazilian race, already mixed with the deformities of the indigenous [Amerindian] peoples and the African." Rather than looking towards Asia to solve any future labour shortage, he felt that Brazil would be strengthened by an infusion of European – specifically northern European – immigrants. This would lead to the whitening of the country's populace at large. Werneck felt Brazil had much to learn from the modernization process underway in the United States where efforts were being made to attract and absorb immigrants. Noting that the two main nationalities – Irish and Germans – who were at the time flowing out of Europe were mainly settling in the United States, Werneck wondered whether Brazil might benefit from the diversion of some of their numbers. On further consideration, Werneck believed that because the Irish had been so long oppressed by the British and had recently also struggled against hunger and starvation, they had acquired an unfortunate tendency towards "turbulent habits" – attitudes that he considered they would be unable to leave behind in Europe, even when offered new and improved opportunities abroad.[11]

It is quite possible that Werneck was alluding to an earlier episode of Irish immigration to Brazil when, in 1826, over 2,400 officers and men (accompanied by several hundred women and children) were recruited in Cork by Colonel William Cotter, an Irish officer serving with the Brazilian army. Many of the recruits signed-up for Brazilian military service in the belief that this would be merely incidental to time spent cultivating the farms that they expected to be granted in Brazil. Instead they were bitterly disappointed by the living and employment conditions that awaited them in Rio de Janeiro and found themselves caught up in the centre of an acrimonious political feud being fought out between the Brazilian emperor, Dom Pedro I, who favoured European immigration, and nativist ministers who displayed a strong antipathy towards foreigners. Dressed in rags, often hungry and subject to brutal treatment by their Brazilian and Italian officers, the Irish were despised by the wider population of the city who taunted them in the streets as being "white slaves". After two years of service, the mercenaries could endure no more: following the savage punishment of one of their number accused of committing a relatively minor offence, they responded by staging a violent, alcohol-fuelled, insurrection, in which the terror they wreaked on the population of Rio de Janeiro could only be put down with the aid of British soldiers.[12] Certainly the memory of the Irish rebellion endured even after Werneck: John Pascoe Grenfell, an officer in the Brazilian navy at the time of the incident, wrote in 1868 while serving as Brazilian consul in Liverpool, that the government would not have any "predilections for the Irish who made a great row when they went out before some years ago".[13]

Other conspicuous failures were those of Welsh and Irish farming settlements that were briefly attempted in the late 1840s and early 1850s in Rio Grande do Sul, Brazil's southernmost province and an area that the Brazilian government was especially keen to populate as a safeguard against possible Argentine encroachment. After a decade of struggle and having received minimal support, most of the already dwindling population of the Welsh colony of Nova Câmbria, located at São Jerônimo, a coal-mining district to the west of the provincial capital, Porto Alegre, were relocated to Patagonia.[14] An Irish settlement at Monte Bonito, in the wooded hills of the Serra das Tapes near Pelotas, also rapidly collapsed. Led by Father T. Donovan, an Irish Catholic priest, many of the three to four hundred immigrants who hailed from County Wexford's Barony of Forth died soon after their arrival. Most of the survivors eventually made their way to Argentina or Uruguay, complaining of the lack of preparations for their reception and of agricultural implements and other tools, of poor land, scarce water and the local diet, especially their having been made to eat "saw dust" – as they considered manioc flour, one of Brazil's staple foods.[15]

Rather than promoting the cause of Irish immigration, Werneck called for the recruitment of German (in particular Protestant) settlers. Werneck was confident that Germans would exert an entirely positive influence on society at large and would pose no threat to social well-being in Imperial Brazil, always being, he believed, "sober, economically sensible, peaceful and hard-working" and carrying with the "a decided love of monarchical institutions". Germans and Irish, he asserted, were thus virtual opposites in terms of their respective characters, demeanours and potential to make positive contributions to the development of the Brazilian nation. "The Irish drink, while the German smoke; the former quarrels, the latter dances and sings", was Werneck's summary of the vices and virtues of the two peoples.[16]

This desire on behalf of the Brazilian elite to pursue Germans as potential immigrants resulted in the publication in Prussia and other German states of a steady stream of books and articles extolling the virtues of Brazil. In truth, the reality was generally very different from the images painted by Brazil's representatives in Europe. The often scandalous conditions experienced by emigrants lured to work on coffee plantations and by the newly installed *colonos* in poorly organized and under-funded land settlement schemes that were generally situated in isolated and semi-lawless parts of Brazil, became a cause for considerable concern and consternation in Germany. To be fair, anti-Brazilian sentiment was encouraged by Hamburg agents representing United States shipping interests, eager to promote their North Atlantic routes and always failing to mention the hardships and dangers that immigrants might similarly encounter in North America. But as a result, by the mid-1850s only the most desperate German emigrants would consider going to Brazil, with those who were tempted being lured by the assisted passages that were on offer. In 1858 the Swiss government recommended that its cantons, a significant source of immigrants for Brazil, issue public cautions against promises being made by recruiters. Soon after, the Sardinian and French governments adopted similar prohibitions. But far greater was the offence in Rio when in July 1859 the Portuguese government ordered that intending emigrants recruited by agents of the Associação Central de Colonisação, an immigration agency closely linked to the Brazilian government, be prevented from departing. A further devastating blow to Brazilian immigration plans was suffered when, in November 1859, the Prussian government, after sporadically struggling to regulate recruiters, issued an edict forbidding emigration to Brazil, a move that was followed by the governments of Baden and Würtemburg and, after the unification of Germany in 1870, by the other states that made up the new country.[17]

Despite these German and Swiss setbacks, the Brazilian government continued to seek the introduction of northern Europeans. The government accepted that immigrants had sometimes been lured to Brazil on contracts only acceptable to "coolies" and was determined to improve the country's image abroad and restore its reputation as a credible immigrant destination.[18] With the experiment of colonization schemes involving independent farmers or sharecroppers making slow headway, and a Brazilian agricultural economy that remained firmly based on large plantations relying on the exploitation of slave labour, proponents of immigration faced a problem. As Germany was now increasingly off-limits as far as Brazilian agents were concerned (although some Germans continued to arrive independently or were recruited in third countries), Brazil was forced to turn its attentions elsewhere to satisfy demands for acceptable settlers. Brazil's representatives abroad made skilful use of local contacts, identifying localities experiencing conditions of particular economic or social unrest whose populations might be the more easily attracted by what was being offered in South America.

During the 1860s and early 1870s, agents representing Brazil scoured the United States and those areas of Europe still open to them, and succeeded in attracting thousands of North Americans, English, Irish, French, Germans from the Russian Volga region, Swedes and Icelanders. Recruitment in Iceland, although hardly a great success, was typical of the efforts being focussed on areas without a tradition of emigration links with South America. A strong emigration sentiment had developed in northern Iceland: a consequence of limited local resources, overpopulation and inequities in land ownership. When relocation to Greenland was put forward, Einar Ásmundsson, an influential local farmer, pointed out that it made little sense to abandon a cold climate only to relocate to one even colder. Apparently well-read, Ásmundsson considered Brazil a much more feasible destination and, with the backing of Brazilian representatives in Europe, a group of four men embarked in 1863 on a journey to South America to explore the possibility of a mass migration. Neighbours waited impatiently for news of the advance party, and by early 1865 some 150 Icelanders had enlisted to join it in Brazil. Although this migration movement evaporated before any actually set off, the idea of Brazil as an attractive destination had been impressed on the local population and in 1873 a further five hundred people declared their intention to leave Iceland. By now Brazilian consular officials in Europe were taking a more active role in the recruitment process and offered to pay the fares of all prospective immigrants. Ultimately, only a few dozen Icelanders were undeterred by the long voyage via Copenhagen and embarked for Brazil, settling near Curitiba.[19]

While always few in number, English and Irish immigrants disembarking in Rio de Janeiro (the main Brazilian port of entry for immigrants) between 1864 and 1873 were only out-numbered by Italians and, by a much more substantial margin, Portuguese. But despite their representing some three-quarters of the immigrant arrivals in Rio, very few Italian or Portuguese immigrants were attracted during these years either to government managed and supported 'state colonies' or to privately-developed agricultural colonies. Most preferred to remain in the Brazilian capital.[20]

To encourage once again the notion of Brazil as being a viable immigrant destination, in 1866 a number of prominent individuals – planters, intellectuals, journalists and politicians – joined forces to establish the Sociedade Internacional da Imigração (International Immigration Society). The Sociedade sought to encourage the Brazilian government to look to

TABLE 1: Immigrant entries to the port of Rio de Janeiro, 1864–73

Year	Portugal	Germany	United States	France	Great Britain	Italy	Spain	Other	Total
1864	5,097	276	106	559	299	872	174	217	7,600
1865	3,784	304	216	534	276	500	170	168	5,952
1866	4,724	244	346	504	416	600	455	192	7,481
1867	4,822	412	1,575	755	867	1,022	280	279	10,012
1868	4,425	563	405	598	1,026	841	218	279	8,355
1869	6,347	376	286	538	375	1,052	332	222	9,528
1870	6,110	306	171	549	427	986	364	210	9,123
1871	8,124	296	191	777	515	1,626	510	292	12,331
1872	12,918	342	219	1,048	1,051	1,808	726	329	18,441
1873	9,907	316	176	852	1,202	1,344	878	256	14,931
Total	66,255	3,435	3,691	6,714	6,454	10,651	4,107	2,444	103,754

Source: Carvalho, *O Brazil: Colonisação e emigração* (1876), p. 497.

immigration as a means to meet the country's future labour requirements. It aimed to arouse public support in favour of immigration and to encourage new legislation to assist would-be immigrants and the positive promotion of the country's image abroad. Port laws, it was argued, should be relaxed to encourage entry points other than Rio de Janeiro, religious toleration should be promoted (aimed primarily at Protestants), encouragement should be given to national groups to settle together in distinct colonies and transportation should be provided to immigrants destined to be employed on plantations or being settled in state agricultural colonies.[21]

In 1866 the Casa de Saúde – also referred to as the 'Emigrants Hotel' or the 'Emigrants Home', a reception centre for newcomers to Brazil – was opened in Rio under the auspices of the Sociedade, intended as a showpiece marking Brazil's ambition of becoming a major immigrant-receiving country. Located on the Morro de Saúde, a hill overlooking central Rio de Janeiro and Guanabara Bay beyond, the hostel was modelled on Castle Garden, New York City's immigrant reception centre. Originally built as a private residence by a wealthy Brazilian, the mansion was later converted for use as a hospital before being turned into a hostel capable, if absolutely necessary, of accommodating as many as five hundred immigrants. By all accounts, conditions there were generally very good: the temporary residents were supplied with comfortable bedding and more than ample food, as well as being given the use of the large and well-tended tropical gardens (the paths shaded by stately imperial palm trees) in which to exercise, relax and recover from their long voyage. A British traveller who visited the hostel in 1870 could only find one slight criticism, that "the place [was] not being kept as clean as it might be", something that he explained was due to the fact that the occupants then were chiefly Irish.[22]

Amongst the most active members of the Sociedade were Brazilians Aurélio Cândido Tavares Bastos, an influential liberal reformer and admirer of Protestantism and the United States, and Quintino Bocaiúva, a journalist and future republican leader. Another key figure was William Scully, the Irish editor and proprietor of the *Anglo-Brazilian Times*, an English-language weekly newspaper published in Rio. Tavares Bastos advocated legal, social, economic and political changes in Brazil that he hoped would stimulate migration from the United States, and through a network of contacts – not least amongst Freemasons – in both countries, he focused his attention on encouraging former Confederates to settle in Brazil.[23] Quintino Bocaiúva would play a central role in the actual dispatch to Brazil of people from North America, though ultimately his selection criteria proved to be extremely elastic, without regard to an applicant's religion, ethnicity, nationality, social or occupational background. Instead, Bocaiúva acted in a manner that was all too typical of emigration agents – that is, his objective was essentially the filling quotas, suggesting that payments to him were based on characteristic 'head money', a fee for each immigrant who was recruited.

Meanwhile, Tipperary-born William Scully sought to encourage the recruitment of immigrants from his native Ireland.[24] Scully considered that the labour market in Rio was generally adequately supplied and that there would be few prospects there or in other Brazilian cities for British clerks, doctors, lawyers, preachers and teachers. Nevertheless, Scully considered that there were some employment possibilities in the city. He believed that Portuguese- and French-speaking British governesses might find openings in Rio, but pointed out that they would require a thorough knowledge of music "as Brazilian ladies possess a talent for it and delight in cultivating it" and also said that "efficient and skilled" female house servants were in demand.[25] For Scully, the real employment possibilities in Brazil were for plantation labourers: he pointed out that he was frequently approached by landed proprietors seeking workers. For those interested, Scully held out the prospect that they might themselves become landowners once their debts for transportation from their home countries had been repaid and after they had gained experience of Brazilian agricultural conditions.[26] Above all he warned "genteel" immigrants to stay away, calling instead for agricultural settlers capable of working hard, able to remain optimistic when faced with the inevitable obstacles and prepared to live for long periods of time on only the most basic of provisions.[27]

In effect Scully's personal mouthpiece, *The Anglo-Brazilian Times* promoted the image of the Brazilian emperor, Pedro II, as an enlightened, benevolent monarch, devoted to modernization, scientific improvement and social welfare, and was generally tactful when taking positions directly hostile to the powerful landed interests and ruling aristocracy. George Lennon Hunt, the British consul in Rio de Janeiro, was scathing of the paper, dismissing it as "*Anglo*" only in name and claiming that it was subsidized and maintained by the Brazilian government for its own purpose. "It is a curious instance of a newspaper with, it may be said, practically only one subscriber", stated Hunt who claimed that the *Times* was not read in Brazil by Englishmen. Instead he considered the paper to be a publicity tool of the Brazilian government, pointing out that its articles were reproduced in the foreign press and translated and published in the local Portuguese-language press, thus receiving wide circulation as coming from an independent source. While there was certainly a considerable element of truth regarding the use made of *The Anglo-Brazilian Times*, the Brazilian government did not escape criticism within its columns. That said, on major political issues of the day the newspaper rarely took positions overtly hostile to the government, for example taking the line on the debate over the abolition of slavery that emancipation should be managed gradually but was inevitable.[28] Scully favoured immigration as a solution to Brazil's labour and land settlement requirements, and through his newspaper advocated immigrants from the United States, England and Ireland, who would bring with them "skills, knowledge, enterprise, and advanced ideas" and infuse into the country "progress, wealth, and empire".[29] Although the paper trod carefully in national politics, generally supportive of whichever party formed the Brazilian government, Scully was often highly critical of immigration policy and practice, not least the failure to provide appropriate levels of financial support. He railed against what he regarded as Brazil's "bigoted marriage laws", arguing that they effectively discriminated against Protestants, those very people from whom he considered the country could most benefit. But while a forceful advocate of immigration, Scully warned against naturalization, owing to the danger of being press-ganged into the Brazilian national guard. Furthermore, he regarded the right to vote as being a "perilous honour", explaining that "if [a citizen] votes he is sure to offend somebody and if he don't vote he offends everybody".[30]

Despite Scully's interest in immigration, he often appeared confused as to whether his aim was to recruit labourers for plantations or independent smallholders to produce essentially for the domestic market. But recognizing that the United States, Canada, Australia, New Zealand and Argentina were all successfully attracting northern European agricultural emigrants, Scully praised what he considered to be Brazil's superior conditions:

Brazil offers to emigrants a climate equalled by few countries, unsurpassed by none. It offers to them, throughout a large extent of country now open to immigration, a climate whose summers neither scorch them with excessive heat nor destroy their crops with drought; and a winter whose cold rarely descends to frost, and not even to that in most districts. It offers to them a climate, and a soil of great richness combined, that permit them, in a temperate clime, to cultivate in security, over a wide spread of region, such valuable and profitable staples as coffee, cotton, tobacco, sugar and other tropicals; to grow a multitude of cereals, fruits and vegetables, both indigenous and of European culture.

It gives them the means of raising on a patch of land, with trifling labor, all the food and edible luxuries needed for the house consumption, and thus, the cattle and horses needing little or nothing, they can find time to devote their main energies to enriching themselves by the cultivation of some small remunerative staple. It offers the poor man with little but his hands the opportunity of acquiring and fitting up a homestead, of supporting himself and improving his condition year by year, without the exhausting labor and exposure requisite in countries scourged with the severe winters of the Northern Continent.[31]

Scully lamented that despite "prospects far superior" to those of other countries, Brazil was only receiving a few thousand immigrants a year, mainly Portuguese and (although far fewer in number) French.[32] He felt, however, that neither nationality made useful contributions to the long-term needs of Brazil. He viewed the Portuguese as unreliable "birds of passage who merely seek to hoe the nests they leave behind", accepting the widely held, but inaccurate, belief that they stayed in Brazil only long enough to accumulate savings to invest back home.[33] French immigrants were also dismissed, the artisans and artists who made up much of their number supposedly pandering in their labours to "luxurious and wasteful tastes", while the peasants amongst them were said to be "unambitious and content with a poverty short of positive starvation".[34] By way of contrast, Scully believed that Germans and British (an inclusive term for Irish as well as English) would be excellent assets as they "emigrate to make their *homes* upon the foreign soil and thus identify themselves with all its present and its future aspirations".[35] On the supposed similarities between the British and German characters, Scully expanded:

> These two nations and their kindred furnish the most desirable immigrants for an undeveloped country. They leave their land of birth to found new homes, and satisfy their longing for a resting place beneath their own roofs, upon their own land, surrounded by their own possessions. They love their native country but they love the house and land they call their own, and, in the second generation, this love of home develops into that love of country that constitutes patriotism. Such immigrants form a veritable yeomanry in the country, and in them lie the greatest elements of its riches and stability of its government, for their productions and their wants give the needed nutriments to commerce and internal trade and such men do not rush madly into revolutions at the call of every disappointed politician, yet they furnish, as we have seen, the finest class for war, if war be needed. Such immigrants it is the interest of every country to attract but such do not seek Brazil, for Brazil does not offer to them the inducements and the facilities for the acquisition of their homesteads which other countries, more appreciative of their value, because wiser and more far-seeing, do not hesitate to furnish.[36]

Scully reasoned that many of the Germans who had not succeeded in Brazil had been the victims of "ill selection" by emigration agents in Europe. Nevertheless, despite generally admiring Germans for their "plodding and industrious" nature, he felt that they would never make real progress in a South American country owing to what he felt was their lack of self-reliance, imagination and enterprise. "In contact with the lower scale of civilization", asserted Scully, quite certain that Brazil was at such a level in the ranks of human achievement, the German, "sinks to or even falls beneath it – meeting a higher he rises but slowly with it". Even in North America's supposedly superior society, Scully believed that it took generations before Germans successfully reached the higher level of achievement of the "Hiberno-Kelt" (whom he believed formed the bed rock of the "North American Anglo-Saxon nation"), but at least they eventually progressed.[37] In contrast to the views expressed a decade earlier by Luiz Werneck, Scully was in no doubt that Brazil could do little better than to look specifically to the Irish as the country's prime immigrant stock:

The Irishman, deeply attached as he is to his own home, his relatives, his friends, abandons useless repinings once he has torn himself away, and seeks to make money, not to return to his former home but to create it anew in his adopted land by bringing those relatives and those friends again around him. Hence the great value of the Irishman as a spontaneous inducer of immigration. The Irishman, perhaps justly accused of unthriftiness and insobriety at home, for he is hopeless there and has the tradition of a bitter oppression to make him feel discontented, becomes active, industrious, and energetic when abroad; intelligent he always is. He soon rids himself of his peculiarities and prejudices and assimilates himself so rapidly with the progressive people around him that his children no longer can be distinguished from the American of centuries of descent, except, perhaps, by a spice of that inherited mother-wit which so eminently qualifies him for a prominent part in the elbowing race of progress. As it is in the United States so it is elsewhere, so it is in the River Plate. Everywhere Irish laborers are in request, for they are appreciated for their intelligence and trustiness.[38]

In an open letter addressed "To the Clergy of Ireland", Scully requested that those sympathetic to the plight of would-be emigrants should screen "families and single men of sober and steady habits" who might be persuaded to consider Brazil as a destination. "In Brazil," wrote Scully, "the Irish emigrant will find kindness shown to him on every side, just as in his native land, and will experience nothing of the unconcealed contempt which the native American is apt to show 'raw' Irishmen." For Catholics, religion, Scully assured his audience, would not be an obstacle affecting an emigrant opting for Brazil, as might be the case in the United States, but he also implied that Irish Protestants would be made just as welcome as Irish Catholics:

> The State Religion of this country is Roman Catholic but all others are fully tolerated. If a sufficient number be settled in any locality it is the usual course of the Brazilian government to provide for them at its cost a clergyman of their communion and country.[39]

The emigrant would have his (and, where appropriate, his family's) passage from Liverpool advanced, the cost (approximately £12) to be deducted from wages after arrival in Brazil, a country that was represented in the usual glowing terms:

The climate [of Brazil] is a healthy one for men not addicted to the abuse of ardent spirits and it is much more agreeable than that of the United States, the heat of summer never reaching the extremes so frequent there, and the winters, in like manner, having in this portion of the Empire much resemblance to an Irish summer.[40]

Any references to aspects of Brazil and Brazilian society that the "Clergy of Ireland" might deem negative – such as the persistence of slavery – were usually conveniently omitted.

Scully's own enthusiasm for Brazil was matched by the country's supporters overseas. His fortnightly newspaper, *The Anglo-Brazilian Times* was widely circulated abroad, its articles reproduced in local journals as an independent source of Brazilian news. The pattern of Arcadian imagery presented by Scully in his *Brazil, Its Provinces and Chief Cities*, a guide published in London in 1866 and 1868, could be clearly seen in books, pamphlets and newspaper articles published in both Britain and the United States, while copies of the book itself were handed out free of charge by Brazilian consulates in England.[41]

Other books and pamphlets were produced in both Britain and the United States for the specific purpose of explaining the superior conditions that Brazil had to offer prospective emigrants. Charles Dunlop's pamphlet *Brazil as a Field for Emigration*, published in London in 1866, was typical of such promotional literature. Brazil – "one of the most magnificent of Empires" – was described by Dunlop as a place "unparalleled for the luxuriance and variety of its vegetable productions", a country which boasted "every description of soil". Reassuring those considering the life of a pioneer in Brazil, the pamphlet declared that "there is little occasion for fear on account of wild beasts" and explained that "the feline species are neither numerous nor formidable, and seldom venture to attack man". Dunlop did, at least, acknowledge that reptiles were unfortunately to be found in Brazil "in an incredible number", but he also advised that, with common-sense precautions, they could safely be avoided. In contrast to this slight risk, emigrants could rest assured that the country's natives were friendly, and that the "manners and customs of the inhabitants are of a genial and conciliatory nature". As for where to choose to settle, Dunlop was in no doubt that the south of the country was most suitable for immigrants arriving from England, in particular recommending Paraná for its "temperate and

1. Agricultural Colonization in Brazil, 1808–67

healthy" climate, and also Santa Catarina where large tracts of land "of the finest quality, and of great richness" were readily available.[42]

One possible drawback of Brazil that was mentioned and then dismissed in guides, was the issue of language. While it was certainly true, conceded Dunlop, that the language of Brazil would be unfamiliar to Englishmen, he explained that Portuguese was "very easy of acquisition", and telling his readers that German immigrants arriving in New York were soon able to cope well with communicating in what he claimed to be the "more complicated" English-language.[43] In his guide book, Scully offered similar reassurances, claiming that in Argentina immigrants found Spanish – "almost identical with Portuguese" – offered little difficulty, and that this language could be picked up in a very short time, "particularly by the Irish".[44]

Travellers' accounts of Brazil were also popular, helping to instil amongst the more literate members of society an image of an exotic country of vast and untapped potential. Published in both Britain and the United States and appearing in numerous editions, *Brazil and the Brazilians* – a book by he American Protestant missionaries James Kidder and Daniel Fletcher – was especially widely circulated. "Let anyone glance at the map of Brazil," exclaimed Kidder and Fletcher, "and he will instantly be convinced that this land is designed by nature for the sustenance of millions." With a "delightful climate [....] refreshed by copious showers" and with no earthquakes or vast deserts, they were convinced that "Providence has designed this land as the residence of a great nation." While the authors were prepared to accept that "it would not be altogether just to compare the administration of law in Brazil to that of England", they considered that nowhere in South America was there to be found "greater personal security and a fairer dispensation of justice".[45] With a measure of British organization, suggested Charles Mansfield, a chemist from Cambridge who travelled through Brazil in the 1850s, "what a Paradise Brazil is, or at least might be, this country if it were possessed by the English!"[46]

Also influential was Thomas Henry Buckle's widely read *History of Civilization in England*. First published in London in 1857 and reprinted many times in subsequent years and decades, the two-volume study sought to explain how the supposedly unique racial and climatic conditions of England led to the emergence of a dynamic civilization. By way of a contrast to the temperate climates of northern Europe,

Buckle included a nine-page description of tropical Brazil. Although he had never actually visited, Buckle described this "land of marvels" in a manner echoing the Garden of Eden:

> Brazil, which is nearly as large as the whole of Europe, is covered with a vegetation of incredible profusion. Indeed, so rank and luxuriant is the growth that Nature seems to riot in the very wantonness of power. A great part of this immense country is filled with dense and tangled forests, whose noble trees, blossoming in unrivalled beauty, and exquisite with a thousand hues, throw out thou produce in endless prodigality. On their summit are perched birds of gorgeous plumage, which nestle in their dark and lofty recesses [in this] vast workshop and repository of Nature.[47]

Brazil, then, was a land that "has always remained entirely uncivilized", its inhabitants mere "wandering savages, incompetent to resist those obstacles which the very bounty of Nature has put in their way". What little civilization that was to be found there after over three centuries of European contact Buckle considered as "very imperfect".[48] But just as Brazilian determinist thinkers were arguing (increasingly influenced by the work of Buckle[49]), British enthusiasts may well have imagined that a 'higher' civilization (here represented by English or Irish) would be capable of reaping Brazil's obvious natural bounty.

Newspapers also frequently accepted uncritically the information offered by proponents of Brazilian immigration, their editors somehow prepared to suspend all disbelief over articles that they would write or insert. "Brazil is a land," enthused *The Universal News*, an Irish Catholic newspaper published in London, "of great natural richness [that is] free from the volcanoes and earthquakes which cause so much terror and devastation in other parts of the South American continent, [while] the healthy winds blowing constantly from the Atlantic keep it free from the droughts and the vicissitudes of season and temperatures which are continually threatening the labour of the agriculturalists in other American states." Its earth, the newspaper assured readers, was "reeking with richness derived from vegetable deposits of countless ages, and the merest scratching of the virgin soil is sufficient to produce crops of embarrassing abundance and highest quality". The economy of this veritable "land of milk and honey" was said to be strong, based on varied exports and the newspaper temptingly pointed out how "gold and diamond mines abound".[50]

To promoters, then, sparsely populated Brazil offered unparalleled opportunities to immigrants of European stock. Brazil offered the conditions necessary to guarantee success for English and Irish emigrants and the country could only benefit from what was firmly believed to be their modernizing influences.

Part III

The Irish believe that they can see an enchanted Island called *O Breasil*, or *O Brasil*, from the isles of Arran; – which General Vallency, in his usual wild way, identifies with the Paradise of Irem.

Robert Southey,
History of Brazil (1810), vol. 1, p. 22.

Brazil is the new land of promise, the new country to be Irishised, as America has been; and if there is but a small percentage of truth in the accounts given of the country, the Irish race will there continue to enjoy the special blessings which the manifest will of the Deity has ever bestowed upon it, albeit that blessing is a reproach in the mouth of the God-belying English – "increase and multiply". Increase and multiply for the Glory of God, and the honour, extension, and strength of the grand old Celtic race. That is the mission of the Brazilian emigrants.

The Universal News, 15 February 1868.

Chapter 2
Brazil and the Irish Diaspora

THE GENERAL BACKGROUND of the mass-migration from Ireland provoked by the Great Famine (1845–6) is well known.[1] The vast majority of the four million people who left Ireland in the decades following the potato blight and the subsequent Famine that in part resulted, crossed the Atlantic to settle in North America, or went in search of employment in the expanding industrial centres of neighbouring England, Scotland and Wales. But others headed for even more distant shores, with Britain's Australian colonies, for example, attracting many tens of thousands of Irish migrants seeking better lives abroad. But despite – or because of – these high numbers, there was also considerable resistance in Ireland to the very idea of emigration. As the historian Kerby Miller explained, emigration was regarded particularly by Irish Catholics as "tantamount to involuntary exile, compelled by British and landlord oppression" and therefore something to be resisted if at all possible.[2] But the Church also feared that emigrants would settle in areas lacking Catholic priests or services, that the settlers and their descendants would soon lose their religion and, in the words of a Dublin journal, "sink into the lowest depths of physical and spiritual degradation".[3] One of the consequences of these attitudes was that remarkably few attempts emerged of colonization schemes aimed specifically at the Irish, and those that were launched were sometimes directed towards Irish already in the United States and not those considering direct emigration.[4]

In pre-Famine Ireland, emigration had posed profound social, cultural, economic, political and psychological problems. But with the onset of the Famine, the steady flow of migrants leaving Ireland developed into a flood. Traditional reservations regarding emigration, while not forgotten, were put aside in the quest to survive the disaster. For the most part, post-Famine Irish emigration followed well-established routes, heading for England or across the Atlantic to New York to scrape a living on the lowest rung of the social and economic ladder.

But for some would-be emigrants, the reassurance of going to a country governed by Catholics made the thought a little less traumatic. Close to home, Spain was briefly promoted in 1861 as a potential destination for Irish emigrants, being offered to people "who might prefer to find a home in the ancient cradle of the Hibernian Milesian race, within two or three days sail from Ireland".[5] Thousands of construction jobs on Spain's expanding railway network were supposedly readily available, and it was also claimed that there was great demand in the country for agricultural labourers. Land too was on offer, but to purchase a smallholding required capital of at least £100 – a substantial sum for most Irish emigrants.[6]

There is no evidence that the supposed Spanish opportunities were in fact pursued. But those Irish desperate to leave their homeland, yet without any savings and lacking family or friends already settled in traditional overseas destinations (who could send them passage money) were vulnerable to schemes that might normally have seemed rashly conceived. One such plan that actually went beyond paper involved a move to create a distinctly Irish-Catholic settlement in French North Africa, a scheme encouraged by Algeria's governor-general, Edmé Patrice MacMahon, himself the son of a Limerick man. In 1869, 130 emigrants – including nineteen families, 28 single men, ten young orphans and a priest – left the port of Queenstown (present-day Cóbh) for North Africa to establish a farming community to be called St Patrice, in honour of the patron saint of Ireland. No sooner had the group arrived in Algeria than *The Cork Examiner* newspaper confidently declared that the venture had "proved a great success". But the reality was very different: just days after their arrival in Algeria, the immigrants were in a state of mutiny, despairing since absolutely no advance preparations had been made for them. In an attempt to bring about an end to the revolt, the perceived ringleaders were expelled from the colony and deported to Marseilles from where they made their way by foot to Paris, hoping to return to Ireland. Within a matter of weeks the rest of the group – who could not face continuing to beg on Algerian streets while waiting for the land, homes and allowances that were still being promised – were shipped off to Marseilles or Bordeaux.[7]

Even further afield, Latin America had long offered tantalizing – and even realistic – opportunities to Irish emigrants of all social backgrounds. Most of the Irish who ventured to the region travelled independently, many labouring on construction projects (especially canals and railways), sometimes enduring slave-like conditions, and, in some instances, having been recruited in the United States where the 'Irish navvy' had long been a feature of the landscape.[8] The Irish also travelled to Latin America as part of settlement schemes, whether formally or informally constituted. A notable feature was an absence of the disapproval that the Catholic Church in Ireland generally displayed to the promotion of migration to English-speaking countries.

One such scheme was focussed on the Mexican frontier province of Texas that, between 1829 and 1834, attracted already relatively well-to-do Irish emigrants. The recruits were neither motivated out of a desperate quest for basic survival, nor were they especially concerned with the issue of the preservation of Irish identity from English oppression. Instead, this pre-Famine current of migration was a carefully thought-out business venture, the participants mainly from small and medium farming families frustrated by Irish laws of primogeniture, high rents and low prices for their produce and who fully intended to make a fortune by investing in land in Mexican Texas. The Mexican government felt that Irish Catholics purchasing land in the colonies of San Patricio and Refugio could be relied on to become loyal citizens and act as a buffer against the southwards encroachment of the United States. Although the Mexican government was ultimately unsuccessful in protecting its borders, the Irish immigrants eventually overcame the hardships of frontier life and divided loyalties, firmly establishing themselves in Texas.[9]

Irish Catholic priests played pivotal roles in the recruitment and settlement of a poorer class of Irish immigrants in Argentina, with the paternalistic efforts from 1844 of Father Anthony Fahy becoming legendary. Thanks to his own knowledge of Argentina and by making use of influential local contacts, over the next few decades Fahy identified the best available sheep-runs, financed the acquisition of flocks of sheep, assisted with marketing and helped secure employment. The socially and economically isolated community was served by itinerant Irish teachers and priests. So successful was Fahy in his endeavours that a distinct Irish Catholic social and religious structure developed in the city and province of Buenos Aires over the following two decades. The Irish church maintained a strong influence over the descendants of the migrants for several generations that endured well into the twentieth century.[10]

Far less well known than these experiences of Mexican Texas and Argentina are those of Irish immigrants elsewhere in Latin America who were unable to overcome the formidable obstacles involved in establishing a self-sustaining community. In partiular one failed attempt may be cited. It involved several hundred Irish men, women and children who – by choice, trickery, hope or desperation – were induced to leave one new homeland (the United States or England) in pursuit of yet another. This was Brazil, a country that the emigrants knew little or nothing about beyond glowing descriptions provided by emigration agents or a parish priest in whom they had invested total trust. What was offered was a Utopian vision of a "New Ireland" in South America that built on Brazilian efforts to attract settlers of northern European extraction. For both the Brazilian promoters of agricultural colonization schemes and for the Irish, so desperate to believe that a better life could be found in far-off South America, the venture ended in abject failure. As part of an immigration process that was ill-conceived, confused in purpose, under-financed and poorly administered, the settlement that received most Irish migrants was also representative of the all-too commonly hopelessly located, physically isolated, maladministered agricultural colonies established in mid-nineteenth century Brazil.

New York City and Wednesbury: Points of Irish Migration

Despite appeals from the likes of William Scully, it was not the downtrodden peasantry in Ireland – whether Catholic or Protestant – that was to emerge as the target of agents representing Brazil. Rather, it was the transplanted Irish living on the fringes of English and United States society who were subjected to inducements to emigrate. Although it is certainly true that not all Irish in mid-nineteenth century England or the United States were living in miserable conditions, a very substantial proportion were, disappointed that their hopes for a better life abroad had amounted to little, if anything. Without any prospects of a return to Ireland, the desperate conditions of the slums of two otherwise very different places – New York City and Wednesbury, an industrial town to the west of Birmingham in the English Midlands – were fertile recruiting grounds for agents, both well-meaning and unscrupulous, who appeared to hold out better prospects in a third country.

The Irish in New York City

The United States had offered an escape route to hundreds of thousands of Irish people fleeing the Famine and the economic and social disruption that followed, but many found that they had exchanged one set of appalling conditions for another. As a result of shipping, family and economic connections, the Famine saw the emergence of New York as the premier port of entry to the United States for Irish immigrants. But even before this mass migration, the New York Irish had suffered more than their share of poverty.[11] Whatever the lure, real or imagined, of New York, the reality was a city as socially and economically divided as any other. For most immigrants in 1860s New York, working, sanitary and housing conditions were horrifying, with the only recently concluded Civil War making it easier for the authorities to neglect the lives of the city's poor. The war had dealt a severe blow to some of New York's traditional industries, such as shipbuilding and the merchant marine, making employment difficult; a situation exacerbated by the arrival of droves of both black and white southerners from the former Confederate states, eager to escape the devastation that surrounded their old homes and anxious for employment.[12]

Physically and psychologically exhausted by their experiences in Ireland, with few occupational skills and with even less capital, whatever their original hopes and dreams may have been, most newly arrived Irish immigrants did not have the capacity to continue beyond New York. One effect of this was to alter drastically the demographics of the city: in 1855 the Irish had made up nearly 30 percent of the population, constituting 44.5 percent of its foreign-born inhabitants. By 1860 New York's 204,000 Irish-born residents represented 24 percent of the population and 53 percent of the foreign-born. Generally speaking, the Irish found employment either as unskilled labourers (in the case of men) or in domestic service (in the case of women) the most menial, insecure and poorly paid occupations on offer.[13]

As might be expected, the living conditions of Irish immigrants in New York reflected the poorly paid and sporadic nature of their work. For a very large proportion, the accommodation that they secured in the cellars and garrets of New York must have felt as squalid as the conditions that they had left behind in Ireland.[14] Their health was frequently poor, hardly surprising considering the poverty and the living conditions that most endured both before and after their arrival in the New World. To add to these physical hardships, anti-Irish feelings ran high and they were often overtly discriminated against in their attempts to secure even the most menial of jobs. The life of an Irish labourer, an emigrant declared in a letter home in 1860, was often "despicable, humiliating, [and] slavish" where there was "no love for him – no protection of life – [he] can be shot down, run through, kicked, cuffed, spat on – and no redress, but a response of served the damn son of an Irish b— right, damn him". Another Irishman commented in the same year that, "The great majority of American people are, in heart and soul anti-Catholic, but more especially anti-Irish [for whom] everything Irish is repugnant to them".[15]

Emigration from New York City to Brazil

Leaving behind the economic devastation that was a consequence of the American Civil War, and to escape what they considered the humiliation and oppression of 'Yankee' rule, thousands of former Confederates chose to 'Reconstruct' themselves abroad. While Europe, Mexico, Venezuela and British Honduras were all significant destinations, it was probably Brazil that attracted the largest number of these voluntary exiles. The promoters of Brazilian agricultural settlement schemes actively courted former Confederates of every background. With land being offered on seemingly attractive terms, in a country being described as possessing unmatched opportunities, it was hardly surprising that thousands of Confederates responded to the call of Brazil. Groups of these southerners settled from the Amazon river port of Santarém in the north of Brazil to Paraná in the south, but only some of those who had been directed to the province of São Paulo succeeded in developing reasonably successful, self-sustaining, communities, of which one or two endured into the twentieth century. A particular attraction of Brazil was that its slave-based plantation economy and society was reassuringly familiar to the white-southern sense of race and class relations. Very few of these Confederate exiles, however, actually purchased slaves in Brazil – indeed immigration and land laws prohibited slavery in any settlements which received aid from the government – but at least they could feel that they were inserting themselves into society that did not condemn their attitudes.[16] "In Brazil", commented a South Carolinian in 1864, there was to be found "a dignity and a hospitality [...] that corresponds in many respects to the lofty and generous bearing which characterized the Southern gentlemen in former times."[17]

While North American leaders of prospective agricultural colonies actively courted their disgruntled compatriots in the southern states, the Brazilian government's recruitment centre was not, as might have been expected, based in New Orleans, Mobile, Savannah or one of the other southern ports. Instead, its centre of operations was New York City – for many of the ruined Confederates one of the most reviled parts of the Union, viewed as the epicentre of Yankee economic domination. Paradoxically though, both New York's authorities and a substantial proportion of its inhabitants had sympathized with the Confederate cause, and the city later attracted substantial numbers of ambitious white southerners.[18] In August 1865, Brazil's consul general in the United States, Luis H. F. de Aguiar, announced the opening of an immigrant recruitment office in New York that would be under the direction of Quintino Bocaiúva, one of the moving forces behind Rio's Sociedade Internacional da Imigração. When in November 1866 Bocaiúva arrived in New York, he took charge of the Brazilian Emigration Agency, installing himself in offices that were located on the then fashionable Broadway, directly across the avenue from the Brazilian Legation. In the intervening period, recruitment had been overseen by the Legation itself, with the Brazilian minister in New York at pains to be as discreet as possible in pursuing his aims so as not to antagonize the United States government by being seen to be tempting its citizens to leave the country or poaching recently arrived European immigrants. On arrival in New York, Bocaiúva joined forces with the New York-based United States and Brazil Steamship Company, owned by Bernardo Caymari, a Cuban with strong business interests in Brazil. In June 1866, Caymari had signed a contract with the Brazilian government to transport emigrants from New York, as it had become clear that most people interested in travelling were without the means to pay their own fares to Rio. The cost of a passage was thirty percent lower than the regular steerage rate, with the Brazilian government agreeing to pay the shipping company three days after the passengers arrived in Rio de Janeiro. The emigrants in turn agreed to reimburse the government in instalments. To reach Brazil, emigrants would first have to travel to New York – a major journey in itself for those living in the southern states – from where they would sail to Rio de Janeiro to begin new lives.[19]

It is certainly true that many, if not most, of those recruited by the Brazilian Emigration Agency were former Confederates. And while some were experienced farmers, many were not: included amongst their number were people from just about every social and economic class of the time, from senior army officers, medical doctors, merchants and ministers of religion to former bounty hunters and barroom loafers with no apparent trade or skill.[20] Other emigrants who were recruited by Bocaiúva were not even former Confederates. Soon after arriving in New York, Bocaiúva announced to the local press a generous package that included passage and land in Brazil on apparently easy terms, measures apparently designed to fill the spare steerage capacity on the Rio de Janeiro-bound ships of the United States and Brazil Steamship Company. Within just a week over one thousand people presented themselves to the Agency's office. To the embarrassment of the Brazilian Legation which was now keen to distance itself from Bocaiúva's rash recruitment drive, around half were destitute Irish and other European immigrants. These men and women were seemingly virtually plucked from the streets of New York, released from jails on the condition that they leave the country or were vagrants taken to the Agency by the local police.[21] Such desperate recruitment methods and aims were not altogether untypical of emigration agents representing Brazilian interests in parts of Europe. Like his official and unofficial counterparts there, Bocaiúva insisted publicly that the Agency would only accept people with agricultural, or related skills, and then only those who could prove their capacity for hard work and sober conduct. The reality was often very different. With payment for services normally in the form of a commission for each person being sent, it would have been surprising had Bocaiúva and Caymari's shipping company not accepted many utterly unsuitable people.[22]

Little is known of the precise operational methods of the New York process. Although some of the recruits were undoubtedly press-ganged into leaving New York, others would have made the decision of their own accord. Given the often appalling living conditions and employment prospects of many people in the city, it is unlikely to have been difficult to find individuals who would have been instantly impressed by the opulence of the Agency's offices and prepared to move south to a country that was being described as somewhere akin to paradise. And while former Confederates wrote most of the Brazilian publicity material circulated at the time in the United States, some addressed a wider audience able to read, or be read to. Typical was a New York-published pamphlet that described in glowing terms the opportunities awaiting immigrants in Brazil, with the province of Santa Catarina – the destination of many of Bocaiúva's recruits – being held out as offering "extensive tracts of land, possessing extraordinary fertility", a mild climate and an excellent network of land and sea communications.[23]

Either displaying unusual levels of critical insight, or perhaps uncomfortable with the image of the United States as a country of emigration, the New York press was sceptical of the recruitment efforts. One editorial comment was that "it is surely a strange thing to see [people] leave this country on so doubtful an invitation, for a land where white labor is not appreciated because of the superabundance of slave labor, and consequently where it is ill-paid".[24] Southern voices, even those normally enthusiastic in their support of Brazilian settlement opportunities, also denounced the profiteering of steamship lines and condemned the Brazilian Emigration Agency's efforts to recruit Yankee and European immigrants in New York, a city where men are "prosperous, victorious, and happy at home".[25] Naturally, Bocaiúva defended the Agency's efforts, taking to task newspaper editors for what he felt was misinformation aimed at damaging the reputation of the Brazilian government.[26] Whether or not people were put off by the voices expressing doubt about the wisdom of the scheme, many did enlist, and the first group of mainly Irish emigrants set off for Brazil at the beginning of 1867.

The Confederate passengers looked upon Bocaiúva's New York Irish with utter contempt. An emigrant from Georgia later described the Irish with whom, in April 1867, he shared the steerage quarters of a Rio de Janeiro-bound steamship, the *North America*, as "scum", accusing the New York authorities of shipping out "the meanest cuss[es] that ever left the country" and of taking advantage of Bocaiúva's apparent willingness to accept anyone who showed up at the Agency's offices.[27]

After a generally gruelling four or five week voyage to Rio, Bocaiúva's recruits were taken to the relative luxury of the Casa de

Saúde, the government's immigration hostel.[28] From there, they were distributed amongst the Ministry of Agriculture's state colonies, places that were quite unprepared for a large influx of settlers despite the official enthusiasm for agricultural settlements. The single largest contingent of immigrants (157 people, of whom thirteen were American, the remainder being English, Irish, Scottish, a few French and Germans), were sent to Porto Alegre, the capital of Rio Grande do Sul. From there, the group was dispersed amongst the province's various, largely German-populated, agricultural settlements, with most sent either to Nova Petrópolis or further north to the remote highlands community of Caseiros (near Lagoa Vermelha).[29] A few of the former New Yorkers were sent to Cananéia, to the south of Santos (in São Paulo province), but it was Colônia Príncipe Dom Pedro, in Santa Catarina, that was allocated by far the largest number, with several substantial parties of Irish origin being sent there, as well as other individuals and small groups. There were also some who remained in Rio while the authorities there attempted to find a place that would accept them.[30]

The Irish in Wednesbury

Irish immigrants were amongst the most socially, economically and politically excluded sectors of the population of nineteenth century England. They have been characterized by the historian Roger Swift as: "[O]utcast from British capitalism as the poorest of the poor, from mainstream British politics as separatist nationalists and republicans, from the 'Anglo-Saxon' race as 'Celts', and as Catholics from the dominant forms of British Protestantism, the Irish were the outcasts of Victorian Britain on the basis of class, nationality, race and religion, with an accumulated body of disadvantages possessed by no other group of similar size, [they were] a people set apart, rejected and despised."[31]

The expansion of industry and the building of the canals and railways created an unparalleled demand for both skilled and unskilled labour in nineteenth century England, with Ireland providing much of the required workforce. As a consequence, 'Little Irelands' emerged in towns and cities throughout Britain, Wednesbury being no exception. Although Wednesbury's Irish community dates from at least the early nineteenth century, it grew substantially in the immediate wake of the Famine. Most of these later arrivals came from Galway and Mayo, densely populated counties in the west of Ireland that remained the most strongly Irish-speaking areas of Ireland even long after the social disruption wrought by the Famine.[32]

If conditions for Irish immigrants in mid-nineteenth England were generally harsh, amongst the worst were those experienced by Black Country communities such as that of Wednesbury, a place referred to as one of the "horrid manufacturing towns" of England.[33] With a population of 21,968 in 1861, Wednesbury was one of a string of towns and villages linked together by chains of metal works and furnaces merging into virtually a single conurbation to form Staffordshire's iron and coal producing district, known as the 'Black Country'. Wednesbury itself had ceased being a mining town and instead was known for the production of axles, girders and wheels and, in particular, iron and brass tubing for locomotive and marine boilers.[34] Described by the American consul in Birmingham as "black by day and red by night",[35] the Black Country impressed observers with the vast concentration of heavy industry that existed within a relatively small area, as well as provoking feelings of horror at the environmental brutality that had been committed across the area. "The landscape, if landscape it can be called," wrote another visitor to the Black Country, "bristles with stunted towers capped with flame, and with tall chimneys vomiting forth clouds of black smoke, which literally roofs the whole region." The soil was contaminated, having long been turned "ink-black" by slurry and other waste, while the air was "hot and stifling and poisoned with mephitic odours". Noise was incessant, with industry creating a deafening "bang and clang and roar and boom of ponderous hammers thundering without the pause of a single moment".[36]

During the 1840s, Wednesbury's approximately three thousand Catholic (overwhelmingly Irish) residents were left virtually unattended. The flood of immigrants arriving in England from Ireland, combined with an

TABLE 2: Population of Wednesbury and the Black Country, 1841–71

	1841	1851	1861	1871
Wednesbury	11,625	14,281	21,968	25,030
Black Country	284,886	362,212	457,329	498,451

Source: Barnsby, *Social Conditions in the Black Country* (1980), pp. 2–3.

increase in the level of self-confidence amongst English Catholics, left the Church stretched far beyond its capacity to meet the spiritual needs of a rapidly growing local Irish Catholic population. To help address this situation, Father George Montgomery was sent in 1850 to Wednesbury to establish a church mission. Montgomery was born in Dublin in 1818 where his father was a Lord Mayor, and grew up in a wealthy and staunchly Protestant family, an unlikely background for one who went on to serve a Catholic community in one of the grimmest corners of industrial England. After taking Holy Orders in the Church of Ireland and a period caring for parishes in Sligo and Dublin, Montgomery became one of numerous Anglican priests to convert to Rome during the 1840s and 1850s. Admitted to Oscott College, a Catholic seminary near Birmingham, Montgomery was ordained in 1849. After a period of study in Rome, he returned to England, preaching to Catholics in Bilston, a Black Country coalmining town, from where he was transferred to neighbouring Wednesbury.[37]

On arrival Montgomery immediately set about raising money to build a church. In this he was successful, with St Mary's Church opening in 1852 on a hilltop overlooking the town.[38] Eager to gain local trust, Montgomery regarded his role as being to serve as the spiritual and moral protector of the town's Catholic – and specifically Irish Catholic – community. Shocked by what he considered to be the miserable and amoral state to which his parishioners had descended, Montgomery saw his duty, as a missionary priest, to play a central role in community life. One of his first crusades was to seek to create order and discipline in the town, in particular to bring a halt to the "deadly melees" that were considered a feature of Wednesbury-Irish life – the police having long dismissed the community as being too "depraved" to make efforts at intervention worthwhile. Montgomery soon won considerable respect and affection from his parishioners and, forever in personal financial debt (the advance on his inheritance spent on church construction) and surviving on the barest of necessities, he was admired, both locally and further afield, for living extremely modestly.[39] Although Montgomery concentrated his efforts on a very local level, his interests took on broader perspectives. In August 1867 he started to produce *The Rev. G. Montgomery's Register*, an occasional publication circulated both within the parish and to friends further afield. The four-page newssheet featured a mix of local church news, passionate declarations by Montgomery concerning political economy and the relationship to poor Catholics in England and extracts from letters that had been received from former parishioners (emigrants living in the United States).[40]

Central to the problems experienced by Wednesbury's Irish residents was the fact that employment was always insecure and living conditions were harsh, even by the brutal standards of mid-nineteenth century industrial England. A petition addressed to Pope Pius IX, drawn up by Montgomery on behalf of 96 Irish "heads of families" explains the community's predicament:

> We support ourselves by manual labour, for most part of the rudest sort, and depend for employment chiefly on the great manufacturing industry of this country. When the trade of England languishes, there is little or no need of our services, and we are frequently altogether deprived of employment for many weeks together. In short, our temporal condition is entirely at the disposal of persons who have no relation to us but that of employers, who, so far as we are concerned, using their money only to make more money, hire us to work, or dismiss us to idleness, as their interests require.[41]

In the periods of economic down-turn which invariably resulted in high levels of unemployment, men were faced with the choice of leaving their families to search for work elsewhere, or abandoning such homes as they had to seek shelter in a dreaded workhouse, maintained by the local Anglican parish, and where religion, implicit in all activity, allowed no concessions to Catholics as non-conformist Protestants at least were granted. While English workers might have experienced similar hardships, the local Irish considered that, for themselves, the consequences were even more socially destructive:

> If we travel about looking for work, we are in danger of departing far from our neighbourhood or priest. If we enter the poor-houses, we go to imprisonment, to forceful companionship with persons not Catholics, who may be hateful to us; we must submit to the yoke of rules which are oppressive, because they were not framed with our consent; and too often are not administered with kindness, but are, in some instances, repugnant to the laws of the Catholic Church. How dreadful the thought that we might die in these places, or that our young children may be immured to them to grow up listless,

faithless parish paupers, having in after-life to struggle for a place in the lowest grade of the social scale, though we had hoped to rear our children to be in all respects better off in this life than we ourselves have been![42]

Even in times when employment was plentiful and well paid, workers suffered from the unrelenting harshness of their work. Irish journalist Hugh Heinrick argued that it was not surprising that so many of the Wednesbury labourers were unable accumulate savings, a consequence of their severe working conditions:

> Too often the wages earned by the severest toil are squandered in the most prodigal folly. The man who labours in the pit, or broils before the furnace from Monday till Saturday – from Saturday till Monday spends the greater part of the wages of the week. It is so with the entire labouring population – English as well as Irish. The prevailing vice here, and, indeed, throughout the whole of the Black Country, is drunkardness [sic]. For this the people endeavor to excuse themselves by the plea that the nature of their occupation is such that drink is indispensable – and that to drink water only would be highly dangerous. Those [...] whose beverage is beer have to take in such quantities to make up for the waste and dryness induced by excessive heat of the places where they work, that their normal condition is that of semi-intoxication. The result in the end is a habit that is most degrading and detrimental to health and morals. To this state too many Irish people descend – and to this, more than to any other cause, is to be attributed the faults and vices by which they are degraded and debased.[43]

Sectarian strife also characterized life in Wednesbury – as it did many other parts of England at the time. The local Anglican church, St James', located in the town's Portway Road district (the heart of Wednesbury's 'Little Ireland') advertised for an Irish-speaking scripture reader in 1851 – less, one supposes, a simple neighbourly gesture of welcome than an attempt to lure converts from amongst the newly arrived Catholics, one small example of denominational rivalry.[44] On a wider front, in the 1860s England's Catholic population was under attack by radical Protestant non-conformists who objected to the Roman Catholic Church being permitted to form dioceses after an enforced interval of three hundred years and who feared that this, combined with the country's increasing Irish population, posed a threat to Protestantism.[45] Politically, too, Irish residents were victimized in retaliation for nationalist activities in both Ireland and England, with the London Irish weekly *The Nation* claiming that, "nowhere in England can our countrymen consider themselves safe from English mob violence".[46] Irish Catholics in Wednesbury felt particularly vulnerable to attacks during periods of economic depression when, perceived as taking away jobs from locals, they could be used as convenient scapegoats,[47] a situation described in Montgomery's petition:

> We are strangers in the land, disliked by the people amongst whom we live, because of our nationality, because of our religion, and because we are in competition with them for employment. During the hours of our work, we have to associate with persons who assail us with blasphemy against the most sacred doctrines of the Catholic religion, with defamation of the clergy and female religious persons, and with obscene discourse. We know also that those around us attribute our poverty, our faults, our follies, and our crimes to the influence of the Catholic faith. This terrible storm of persecution, of calumny, is sufficient to overwhelm persons more steadfast than we are; and we tremble when we think of the effect which it has on our children. At all times this tempest is felt by the poor Irish in England, but just now – excited by the fraud and malice of certain fanatics and apostates – it rages with fury against us.[48]

It is against this background of poverty, insecurity, religious and ethnic strife that Wednesbury proved to be a fertile recruiting ground for agents seeking emigrants for an unfamiliar and distant part of the world.

Emigration from Wednesbury to Brazil

As the Wednesbury mission became secure, Montgomery concentrated his attention on education and emigration. Montgomery was convinced that the British state was thoroughly untrustworthy and possessed by an irreconcilable hatred of the Catholic religion. Certain that the state's true purpose behind the subsidizing of Catholic schools was to exert control over the Church through financial means, Montgomery called for self-reliance, urging priests and laity to establish and maintain schools on a strictly independent basis and set an example with the Wednesbury mission school. But it was to emigration that Montgomery dedicated much of his energy. He argued that agriculture, rather than manufacturing or industry, was the more "eligible" way of life, and was convinced that "as God had given the earth to the children of men", it was the necessary work of both "enlightened statesmanship" and "Christian Charity" to assist families of destitute workers to migrate overseas where they could take possession of uninhabited fertile lands that were crying out for exploitation.[49]

Soon after taking up his position in Wednesbury, Montgomery began receiving letters from Irish former residents of the town who had emigrated to the United States, hundreds of whom had settled in New York, Ohio and Pennsylvania. These letters frequently contained passages paid for emigrants' friends and relations who had been left behind in Wednesbury, causing Montgomery to observe that his mission was in effect serving as a depot for United States-bound emigrants.[50] Recognizing this reality, Montgomery felt justified to intervene directly in the migration process, taking it upon himself to investigate possible new destinations and enter into negotiations with their agents. Indeed, given the conditions that prevailed in Wednesbury, not only did he feel that it was appropriate to assist his parishioners to emigrate, he felt that it his religious duty to do so, declaring: "We hear our divine Saviour saying, *'When they persecute you in one state, flee ye to another,*' and we look whither we may flee to obey this precept."[51] Montgomery first considered an Oregon settlement scheme, and in 1853 unsuccessfully attempted to raise funds to visit the United States where he hoped to find wealthy Irish-American patrons willing to finance agricultural settlements in the West. Of the motives behind this plan, Montgomery later recalled, "it seemed to me a pity that the expatriated Catholic peasants of Ireland should die out in the English towns – a miserable proletarian population without religion or patriotism".[52]

Montgomery argued that if the Irish were to remain in England, it was vital that they improve their position economically as "without temporal prosperity – speaking of the run of mankind, and taking people in masses – there can be no *spiritual* prosperity".[53] He felt, however, that even a modest standard of living in England was an unrealistic goal, with the best that he might achieve for his parishioners being "to dress the wounds of the perishing wayfarer". For there to be a hope of eternal salvation, Montgomery concluded that the Irish must escape England, to be "conveyed to a place where [they] may be thoroughly taken care of".[54] Acknowledging, however, the Church's ambivalent attitude with regards emigration from Ireland, Montgomery was at pains to point out that the situation of the Irish in England was entirely different:

> I am not disturbing a people who are at home contented and settled, but I am trying to direct their migrations people who are on the move in search of a home. To my view the Irish in England, considered as a body, are like the traveller in the Gospel, who lay in the way 'stripped and wounded and half dead'. The poor people are wounded with five grievous wounds. They are suffering compulsory and extreme poverty; they are strangers in the land; they are expatriated strangers, who have neither country nor home; their progeny is becoming extinct in the cities and great towns of England; and their children are apostatising from the Catholic faith.[55]

Although Montgomery was prepared to accept that the spiritual condition of Catholics in the United States was slightly better than was the case of those in England, he lamented the danger to faith and morals that he believed Catholics continuously faced in both of these Protestant-dominated countries.[56] Montgomery would most likely have agreed with Heinrick's comment that "[n]early all that is good in our people here is Irish. Three-fourth of what is bad is acquired – and acquired from the habits, manners, and morals of the lowest and worst among the English people."[57] Due to the many dangers and influences in both England and the United States, Montgomery was instead keen to encourage migration to a Catholic country, one where the Irish would enjoy protection, security of faith and morals; impossible, agreed Henry Formby, a fellow Catholic priest and admirer of Montgomery, either in England or in "the mixed and often godless society of the United States".[58] Montgomery considered Brazil to be the ideal destination for poor Catholic emigrants, for "in its immense territory and its teeming soil, [Brazil] has the means to sustain twenty times its present population, and in its wise Ruler and generous people has the will to obey the admonition of the prophet, 'Educ vagos et egenos in domun tuum,' – *bring the needy and the harbourless*; and to merit to hear from our Lord, 'Hospes eram et collegistis me,' – *I was a stranger and you took me in*." Montgomery praised the Brazilian emperor as being a compassionate and truly enlightened leader who was offering "the poorest strangers a welcome such as a nation never gave, and promises to extend over his humble guests a fostering care such as a Prince has never had for his people".[59] Montgomery felt that it was imperative that nothing should get in the way of dispatching as many emigrants as possible to Brazil, going so far as to declare that "whoever shall attempt to obstruct the movement which aims at settling hardy, virtuous, industrious, poor Catholics as agriculturalists in Brazil, will be an enemy of the human race."[60]

It is not exactly clear how Montgomery became such a fervent proponent of Brazil but he was certainly well-read and may well have been influenced by the colourful Brazilian passages in Thomas Henry Buckle's *History of Civilization in England*.[61] Montgomery himself reported that he began to seriously consider the practical possibility of Brazil as a destination for Catholic emigrants from the British Isles in 1866 after reading an article in the *Standard*, a newspaper published in London. "In no latitude," the article extolled in language distinctly reminiscent of that in William Scully's newly published guidebook to Brazil and Charles Dunlop's pamphlet promoting emigration to Brazil – indeed, also bearing a striking similarity to that used by Thomas Buckle – "can there be discovered greater national wealth. The surface is enormous, the soil exuberant, the seaports are magnificent, the navigable rivers unparalleled, the mines inexhaustible; and yet Brazil pines for people."[62] With such a country clearly yearning for immigrants, Montgomery decided that it was his duty to find such people. After reading the *Standard* article, Montgomery entered into correspondence with its author, said to be an Englishman who had lived in Brazil for fifteen years. Encouraged by all that he heard, Montgomery went on to canvass the opinions of others with first-hand experience of the country. Having satisfied himself that Brazil was definitely "a fit place for the settlement of poor Catholics astray in England", Montgomery began to take practical measures to assist his parishioners to emigrate.[63]

In November 1867, Joaquim Maria de Almeida Portugal, a former officer in the Brazilian navy who presented himself as the "London representative" of his self-styled Commercial Agency of Brazil, placed his first advertisements in English papers promoting Brazilian emigration:[64]

EMIGRATION TO THE BRAZILS
The most PRODUCTIVE and PROMISING FIELD upon the GLOBE.
ASSISTED and FREE PASSAGES
from England to the Ports of Brazil.
BEDDING, COOKING and all SHIPPING UTENSILS allowed free.
GRANTS OF LAND and facilities afforded for Permanent Settlement.
EMIGRANTS TOOLS, AGRICULTURAL IMPLEMENTS
AND MACHINERY
FREE OF TRANSPORT.
Emigrants' Home open and Guide provided.
LABOUR IN DEMAND AND GOOD WAGES.
Apply to Le Chevalier J. de Almeida Portugal
COMMERCIAL AGENCY OF BRAZIL
Westminster Chambers, Victoria Street, London[65]

The terms that Almeida Portugal was offering were, in the opinion of a British Colonial Office emigration commissioner, "very liberal and advantageous", especially that of the free and assisted passages (meaning an advance of the ticket with a grant representing the difference in cost between the lower steerage fare from Liverpool to New York and the more distant and less travelled Liverpool to Rio de Janeiro route) and "grants of land" – though the interpretation of this term was to be later a point of contention.[66]

With Brazil now being offered as an alternative and an apparently attractive and realistic destination, in 1867 Montgomery approached Father Joseph Lazenby, an Irish Jesuit at the Colégio do Santissimo Salvador (the College of the Most Holy Saviour) in Desterro, the capital of the province of Santa Catarina, for advice on local conditions.[67] Lazenby urged Montgomery to encourage his parishioners to emigrate to Brazil and appealed to him for help to recruit a priest "who can manage the Irish" already in Príncipe Dom Pedro, saying that the government would provide a house, stable, chapel and school along with an annual payment of 800 milreis (£83 6s.) for the chaplain and a further 360 milreis (£37 10s.) towards the costs of purchasing and maintaining a horse.[68] This news that there were Irish already settled in Santa Catarina clearly delighted Montgomery who took this simple fact as evidence that Brazil must be a suitable place to dispatch his own Irish followers.[69]

Excited by the opportunity that presented itself, Montgomery sought the approval of the Bishop of Birmingham to take up the position of chaplain. Permission was granted, but on the condition that Montgomery should first clear all his debts – debts that he insisted had "been occasioned by my efforts to promote the welfare of my poor Catholic countrymen in whose service I have spent my private fortune". Eager to travel to Brazil as soon as possible, Montgomery appealed to his parishioners and to his supporters further afield for donations to enable him to escape what he described as a "hateful place", pleading that "a very little from each one of the many interested in my proceedings will accomplish all that I require; let it be given as it is asked, in the name of God and our Lady."[70] Lazenby, who had taken a keen interest in the arrival of the New York Irish in 1867, went to Príncipe Dom Pedro shortly before Christmas of that year to serve as acting chaplain pending the arrival of Montgomery, urging him to finalize arrangements as soon as possible.[71]

Quite possibly after having seen the advertisements placed in the name of the Commercial Agency of Brazil, Montgomery made contact with Joaquim de Almeida Portugal who was now working with the immigration agents Meadows and Christopher, most of whose recruitment activities were concentrated in London. Montgomery was much taken by Almeida Portugal, whom he described as "a devout Catholic, a polished gentleman, and a warm friend of the Irish people," and they quickly set about the practicalities of realizing the priest's long-held dream of organizing and dispatching a party of emigrants. Word rapidly spread in Wednesbury and beyond of Montgomery's ideas and intentions of assisting Catholics to emigrate. While insisting that he was not personally an agent, Montgomery claimed that he was being "deluged" with enquiries about Brazil – so many that he decided to form, in the first instance, two groups drawn only from Wednesbury. Montgomery would organize still further departures and said that he was planning to accompany the second group to Brazil where he would remain to receive further groups of immigrant arriving from England.[72]

At five o'clock in the wintry morning of Monday, 3 February 1868, hundreds of warmly clad Brazil-bound emigrants began assembling at Wednesbury's London and North Western Railway Station. The previous day Montgomery had delivered a moving valedictory sermon at St Mary's Church after which the emigrants completed packing their

A Black Country townscape. (British Library)

belongings before spending their final hours in Wednesbury with the friends they were leaving behind. By seven-thirty on the morning of departure, the railway station platform was packed with emigrants and their well-wishers, and adjacent platforms and bridges were lined with thousands of spectators. There were men, grimy and sweaty after a night labouring amidst the intense heat of an iron foundry and others, bright from the weekly bath of the previous day and stopping off on their way to begin the working week. There were women and children, straight out of bed with just a cape thrown over them, all eager to wave goodbye and share the excitement of departing friends. Also present to bid farewell were Meadows, a Mr Louis (there to represent Almeida Portugal) and Montgomery, who was eagerly anticipating joining his emigrating parishioners in Brazil as soon as possible. The general mood was one of cheerful excitement though, having had little sleep the previous night, the children appeared distressed by the hubbub. After "parting glasses" of spirits were raised and emptied, the 339 emigrants (256 Irish, 75 English, four Scottish, three Dutch and one of undetermined nationality who died during the sea voyage) and their baggage were loaded into the seven carriages of the train that had been reserved for their use. Of the group, 247 people were recruited in Wednesbury by Montgomery on behalf of Joaquim de Almeida Portugal, while the other 92 – also mainly Catholics – were recruited mainly in Birmingham and London by the emigration agents purporting to represent the Brazilian government. At seven forty-five the train slowly pulled out of the station amidst cheering and the waving of hats and handkerchiefs by the assembled onlookers. The general mood was one of celebration rather than sorrow, with far fewer tears shed than normally seen at Black Country stations when groups of just half-a-dozen or so people would set off to begin new lives overseas. In contrast, the passage money that was advanced to Brazil-bound emigrants allowed for entire extended families to travel together, not leaving close relatives behind to mourn their departure, while those friends who were remaining in Wednesbury fully expected that they would be embarking soon for South America.[73]

From Wednesbury, the group travelled northwest to Liverpool where they had assumed they would board a ship for Brazil that same evening. Both the emigrants and Montgomery were unaware, however, that Almeida Portugal's legal authority to recruit on behalf of Brazil was being fiercely challenged, leading to delays, confusion and discord that boded ill for their future prospects in their new home.

The group was due to sail to Rio de Janeiro on the *Florence Chipman*, a steamship chartered from John S. Wolf and Company by the Liverpool shipping agents Kitto, Woodward and Clarke on behalf of Joaquim de Almeida Portugal. The shipping agents had accepted payment by bills of credit to the value of £3,700 – with a guarantee of a further £800 on delivery of the emigrants to Rio – issued by Almeida Portugal acting on behalf, so he claimed, of the Brazilian government. In spite of Almeida Portugal's claimed links, the Brazilian consulates in Liverpool and London refused to honour these promissory notes. Washing their hands of any responsibility, the Brazilian Legation in London explained that Almeida Portugal "neither was ever, nor is he now, employed to enter into contracts for Emigration from the United Kingdom – the Brazilian government has no knowledge whatsoever of any contract or contracts entered into by Almeida Portugal and [it] has therefore not repudiated such contracts".[74] The Legation accepted that in May 1867 Almeida Portugal had offered his services to promote emigration from Ireland to Brazil and that in July the Brazilian minister of agriculture, Manoel Pinto de Souza Dantas, had appointed him to act under the supervision of the consul in Liverpool, but insisted that actual authority to engage emigrants was never granted. As early as December 1867, the Ministry of Agriculture issued denials that Almeida Portugal and his Commercial Agency of Brazil had authority to recruit prospective emigrants, reminding people that grants of land were not permitted under Brazilian law, with publicly owned land being only available by way of purchase, albeit with the benefit of lengthy periods of credit.[75] The Legation pointed out that Almeida Portugal's advertisements contained "statements of such extraordinary character as to Brazilian emigration and so contrary to Brazilian Legislation on the subject" that on 14 January 1868 he had been informed in writing that he should suspend all his recruitment activities. As far as the Wednesbury party was concerned, while expressing great sympathy with the plight of the "unfortunate emigrants" who were stranded in Liverpool, the Legation insisted that it was powerless to intervene to help them on their way.[76]

In his defence, Almeida Portugal vehemently denied that he had ever claimed that land was available without payment in Brazil – simply that "grants of land" were available. Nor, he insisted, was he claiming that his Commercial Agency of Brazil had any direct government connection beyond the authority to act as an emigration agent in England. While it is certainly correct that Almeida Portugal's earlier advertisements were, whether deliberately or not, ambiguous, the wording in later ones (possibly placed to cover earlier errors or misrepresentations) was clearer. Offers of "free and assisted passages" were made and the "sale of tracts of superior land suitable for agricultural and pastoral purposes" was specified, terms that the Wednesbury party seemed to fully understand.[77]

When news reached Brazil of the arguments raging in England, the journalist and Irish immigration advocate William Scully directed his anger at Almeida Portugal and his "misrepresentations". That said, he also expressed a hope that the episode would at least have the beneficial effect of clarifying who abroad was entrusted with powers to assist or encourage emigration to Brazil.[78] But Almeida Portugal himself took little part in these arguments and none at all in attempting to convince Brazil's official representatives in England to accept responsibility for the marooned emigrants who were being sheltered at the Emigration Depot in Birkenhead (across the Mersey Estuary from Liverpool), normally reserved only for emigrants en route to Empire destinations. In late January Almeida Portugal returned to Brazil, either to escape the wrath of the frustrated would-be emigrants and the fury of those holding worthless bills of credit or, more innocently, to be sure to reach Rio de Janeiro in advance of his recruits.

Hearing of the arguments over the responsibility for the emigrants seemingly abandoned in Birkenhead, Scully expressed amazement for the "gullibility and unbusinesslike carelessness" of those in England who had accepted worthless bills of credit, and not heeded his advice that the Brazilian government had not appointed an agent in the British Isles authorized to provide assisted immigration. "The mercantile firms", commented Scully not without a degree of smugness, "who permitted themselves so easily to be hoodwinked by plausible but baseless representations have not a prima facie or even a colorable plea on their behalf, and, if the government has determined to meet the requisitions made upon it on the behalf of the charterers of the *Florence Chipman*, it is not because they have the slightest claim upon it or because its honor is in the smallest degree involved, but solely because of the disagreeable and unfortunate position in which the innocent immigrants themselves would be placed by any other settlement of this imbroglio." Trying at least to find something positive to report regarding the utter confusion that the *Florence Chipman* affair had turned into, Scully at least felt that Brazil's potential for obtaining immigrants in Britain had been clearly demonstrated.[79]

While emigrants waited in Birkenhead for a decision on their fate, the British government's Emigration Commission, the Foreign Office and the shipping agents Kitto, Woodward and Clarke all tried to convince the Brazilian Legation to accept responsibility for the commitments made in its government's name.[80] As the delay in the departure lengthened, discontent mounted amongst the would-be emigrants receiving emergency shelter in Birkenhead. With nothing to do but wait, the Wednesbury party easily absorbed rumours that always rapidly spread within the Liverpool and Birkenhead emigrant community. One such rumour that reached the Emigration Depot was the suggestion that the true purpose behind the recruitment was to supply recruits to bolster the Brazilian forces in the execution of the Paraguayan War – the brutal War of the Triple Alliance, 1865–70 – being waged by Brazil and its allies Argentina and Uruguay against Paraguay. The emigrants' supporters dismissed such suggestions, pointing out that on board the *Florence Chipman* were agricultural machines suitable for "the hard but prolific soils of Rio Grande".[81] These concerns, however, held the possibility of being legitimate: since Independence South American armies and navies (not least Brazil's) had often recruited into their ranks British and Irish soldiers and sailors, enticed by offers of high wages or even land grants. Furthermore, many German immigrants in Dona Francisca (Joinville), Blumenau, Brusque and elsewhere in Santa Catarina were pressed to enlist as so-called Voluntários da Pátria to fight against Paraguay (as there was no official compulsory military service in Brazil) with many sacrificing their lives. The allegations, therefore, were perfectly plausible – indeed, by the end of the year, at least one party of such mercenaries (a total of 230 men) travelled out clandestinely from England to join the Brazilian army, attracted by a promised £50 bounty to be paid on arrival.[82]

Despite the anxieties surrounding the delay, John Connolly (Montgomery's unofficial representative travelling with the Wednesbury emigrants) felt able to report back that most of Montgomery's recruits were "well, hearty, and contented", assuring the priest that the litany was being recited, and that night prayers and the rosary were said daily. Even so, Connolly acknowledged that some Wednesbury people were unhappy, although these were dismissed merely as "a few women, who, I am afraid, you seldom saw at Mass, judging from their language and quarrelsome disposition". The Birmingham emigrants, however, were said to be more troublesome, described as " great fault-finders and find fault with [Montgomery] more than anyone else for sending them here", with one seeking to return home "to tell the Bishop how you have sent them to such misery".[83]

Meanwhile, Montgomery's supporters continued to promote his mission, reminding all who would listen of the good fortune that the party – and future groups of emigrants – could expect. The sympathetic *Universal News* was confident that Brazil offered all the geographical ingredients required for agricultural success. But possibly most encouraging to this strident Irish Catholic voice was that Catholicism was Brazil's state religion. In contrast with the experience in England, it was stated that in Brazil, "Irishmen will be free from the outrage of paying for the support of a hostile Church of a miserable and selfish minority". Like Dunlop's pamphlet, *The Universal News* acknowledged that the language of Brazil would be alien, but the paper reminded its readers that for Irish people the English language was also an adopted one, an imposed language that was "a symbol of oppression and humiliation". In South America rested the hope of an Irish cultural renaissance, with *The Universal News* saying that just as the Welsh language was being implanted in Patagonia, responsibility now "rests with [the Irish emigrants] to make the Irish language – through their schools and through their newspapers – the future language of their colony". Sparsely populated Brazil thus presented an ideal environment for the creation of what was expected to be the first "New Ireland" in South America.[84]

After a delay of almost three weeks, the British government's Emigration Commission and Brazilian diplomatic representatives in England jointly agreed to pay to transport the group to Rio de Janeiro. Brazil's Liverpool consul, Admiral John Pascoe Grenfell, issued the emigrants with a joint passport and on Saturday 22 February 1868 the *Florence Chipman* set sail for South America.[85] Supporters of the scheme seemed immediately to have forgotten recent arguments and were utterly oblivious to the implications that the chaos might hold for the emigrants' future welfare. On the contrary, *The Universal News* praised what was occurring as "a most important change in the character of Irish emigration, and one that may greatly influence the future of the Irish race". Despite the set-backs, it was announced that the *Florence Chipman* emigrants represented the first of twenty-thousand people who would soon be leaving Britain for agricultural and mining districts of Brazil, of whom as many as eight thousand would depart by the year's end.[86]

Montgomery was especially satisfied at the departure and turned his thoughts to building on what he confidently believed to be a successful start to his venture. No sooner had the *Florence Chipman* set off, than Montgomery announced plans to establish an emigration association, to be called 'The Pioneers of Our Lady of Help', the aim of which would be to settle Irish Catholics in what were assumed to be unoccupied forest lands in South America, first in Brazil and then in neighbouring countries. As the intention was to construct self-sustaining communities – a network of "New Irelands" – people from all walks of life were urged to come forward. Priests, medical doctors, engineers, military men, forestry experts, naturalists and taxidermists (as well as "single men and lads, free, healthy, brave, strong, generous, and disposed to live as becomes good Catholics") would all be needed to construct the new Irish societies that were to emerge in South America. The host countries, Montgomery explained, would provide employment and salaries to the immigrants while they settled into their new lives and as roads were constructed opening up the virgin forest for exploitation and development.[87] Although no other priests were so far directly involved, Montgomery expected that "scores" of Catholic clergy would soon volunteer their services to assist in the development of "New Irelands" in South America, men who would provide all-important spiritual leadership during the early years in a sparsely populated Brazil rather than merely follow in the footsteps of other emigrants as he felt was all too typical of the pattern in North America.[88]

Chapter 3
A "New Ireland" in Brazil

FORTUNATELY FOR THE Black Country emigrants, their Atlantic crossing was largely an uneventful one. There was little to break the tedium of the two-month voyage apart from watching out for dolphins that often sailed alongside the ship and commenting on the ever-changing temperature as the vessel approached the tropics. Although mid-nineteenth century emigrant ships carried with them fearful reputations, the "roomy, well ventilated and lighted" *Florence Chipman* was more comfortable than many such vessels. Perhaps because the British government's emigration commissioners more closely supervised the ship's departure than was common for other independently-owned and chartered vessels leaving Liverpool, the *Florence Chipman* was particularly well stocked with provisions for the long voyage ahead. For example, supplies of meat on board were so plentiful as to cause John Connolly to comment that they were entirely wasted on "those who never saw beef but once a fortnight".[1]

The *Florence Chipman* sailed into Rio de Janeiro's Guanabara Bay on 22 April 1868, laying anchor just offshore from the centre of the city.[2] Landing at the Mauá wharf, the immigrants' first impression of Brazil was unlikely to have been a positive one. The area was described by Richard Burton, the British diplomat and explorer, as being as squalid as the very worst parts of London's docklands.[3] Fortunately, though, the wharf was only a short walk below the Casa de Saúde, the secure and peaceful immigrants' hostel where the ship's passengers were to be lodged.

As news spread across the city of the arrival of the *Florence Chipman* and its cargo of several hundred British subjects, curious Brazilian and British residents of Rio made arrangements to visit the hostel. Amongst the first to call was the emperor, Pedro II, who visited the hostel on 24 April, accompanied by a large retinue that included the minister of agriculture and the official agent of colonization. Also in attendance was the until recently disowned Joaquim de Almeida Portugal – his presence alongside Dom Pedro more than suggesting that his mission to England had had, as he always insisted, at least some official blessing. To mark the immigrants' safe arrival in Brazil, a mass of thanksgiving was celebrated in the chapel that adjoined the Casa de Saúde, after which Dom Pedro and his retinue of courtiers spent two hours talking to the new arrivals, inspecting their accommodation and food.[4] "His visit", reported *The Anglo-Brazilian Times*, "was greatly appreciated by the immigrants and both on his coming and his departure he was heartily cheered by them." In turn, the emperor was said to have been, "greatly pleased with the quiet and respectable appearance of the emigrants and especially with the blooming faces of the children" – of whom there were 146 under twelve-years of age.[5]

Another early visitor to the hostel was William Scully, the editor of *The Anglo-Brazilian Times*, who surprised himself by being impressed by the high quality of the newly arrived immigrants. Congratulating Almeida Portugal on the make-up of the group that he had assembled in England, as well as Brazil on its good fortune, a much-relieved Scully boasted that the new arrivals were, "the proper kind [of immigrants] and precisely the sort we proposed that the Brazilian government would import instead of wasting the money on bringing in vagabonds and scamps, as most of those from New York were". While acknowledging that his newspaper had earlier condemned the Black Country recruitment efforts, after being informed that they were going on without Brazilian government approval, Scully now felt that Almeida Portugal's actions deserved appreciation and public congratulation – whatever the precise circumstances.[6]

Although the likes of Dom Pedro and Scully had clear reasons for wanting to be convinced of the quality of the immigrants, the group gave even hardened sceptics pause for thought. One such individual was Johann Jacob Sturz, a Prussian diplomat with a record of being highly critical of Brazilian immigration and colonization policies and practices. Sturz initially considered the very idea of importing Irish immigrants from England to be utterly farcical; after meeting them in Rio he was prepared to concede that despite disappointment (although hardly much surprise) that there were few agriculturalists amongst the Wednesbury recruits, he felt that they were "a better type of people that I had expected, not at all riff-raff and not overly drunken".[7]

Reports reached Wednesbury of the group's enthusiastic reception in Rio. "The emigrants have been received in hospitable Brazil with such a welcome as poor Irish people never had on earth", wrote an obviously delighted and presumably – given the circumstances of the departure from Birkenhead – relieved Montgomery. "Hurrah for Dom Pedro, the second Emperor of Brazil! Long may he live, and long may he reign. And blessings, too, to be in the land of Chevalier Joaquim de Almeida Portugal for his kindness to the Irish poor." The future looked promising to Montgomery: he could now confidently concentrate on recruiting a second party of emigrants, to be soon followed across the Atlantic by what he anticipated would be half a million "virtuous and industrious" Irish and English Catholics who would make Brazil, or some other part of South America, their new homeland.[8]

The immigrants were content with the living conditions in the palatial Casa de Saúde – not least when compared to the overcrowded Emigrants Depot in Birkenhead, a place intended for stays of no more than a few days rather than the several weeks that had been forced on them. The delay in departing England turned out to be fortuitous as it meant that the emigrants avoided the worst of the Brazilian summer heat, rain and the seasonal outbreaks of yellow fever. But there were some immediate problems, in particular the children finding it difficult to tolerate an unfamiliar diet centred on large rations of beef. The management of the Casa de Saúde – the showpiece of Brazil's immigration program which consistently succeeded in maintaining excellent standards – was responsive to this concern, and supplied the children with familiar milk and bread, until then not included amongst the rations. A far more difficult problem for the authorities, however, was what to do with hundreds of newly arrived immigrants who had expected that they would be transferred immediately to agricultural colonies. Although their impending arrival in Rio de Janeiro must have been known for weeks, the Ministry of Agriculture had failed to make advance arrangements for any of the state colonies to receive the immigrants. Due to the delay, and to offset the expense of providing the immigrants with temporary board and lodging, both men and women were encouraged to accept offers of work. Women were particularly in demand, receiving offers of employment as domestic servants with British families living in Rio. Only two or three women accepted positions as a housemaid or cook, and a similar number of the young men or boys who were offered jobs decided to accept them. It was the promise of land that had drawn the immigrants to Brazil: they were not going to be so easily tempted by offers of jobs little or no better than those that had been available to them in England.[9]

Even before the *Florence Chipman*'s arrival in Rio, Scully had been urging caution when considering where to place immigrants for agricultural settlement, in particular warning against the suitability of the "timber lands of Santa Catarina". Aware of the difficulties experienced by some immigrants recently sent to Santa Catarina, Scully argued that the province had little to favour it, not least a climate that he considered inappropriate for agricultural settlement, and described as "an unruly compromise between two contending climes, it has neither the advantages of the steady tropic heats nor the invigorating coolness of the uplands". In Scully's opinion, it was positively inhumane to place northern Europeans "in the close valleys and the dense impenetrable thickets of the woodlands" of Santa Catarina, conditions that were only suited to "the Portuguese of the islands [the Azores] and to the negroes of the planters, reared to the use of the foice and hoe alone". Immigrants from "the cool and open north", he believed, had different requirements to have a chance of succeeding in Brazil. "Accustomed to labor with the plough and the scythe," wrote Scully of the northern European, "to having domestic animals around him to assist him and to furnish his children's food, nothing is so apt to create home sickness than being thus condemned to virtual isolation in the woody thicket, to the incessant and monotonous labor of the foice [blade] and hoe, to the need of transporting every necessary on his back though narrow paths: it is simplifying life which only habit long continued can make acceptable, however fertile be the lands." Instead, Scully urged that prairie land be provided in either Paraná or Rio Grande do Sul, a condition that he regarded as "*an absolute necessity*, if English and Irish immigration is to prove satisfactory and bring about that spontaneous immigration which will spring up if the settlements are successes". Scully specifically recommended the region around Curitiba, an area that he felt showed great potential for agricultural development, notwithstanding its poor soils.[10]

But with absolutely no understanding or knowledge of Brazilian regional conditions, the priority for the *Florence Chipman* immigrants was to avoid being disbanded as a group, a position that the tightly-knit

Wednesbury group held as especially important. Only the Colônia Príncipe Dom Pedro was deemed by the Brazilian government's agent for colonization as being capable of absorbing a substantial contingent of immigrants. Because of this, two hundred of the immigrants – including most of the Wednesbury contingent – soon boarded a coastal steamer for Santa Catarina.[11] Apart from the few men and women who had accepted offers of work in Rio, most of the *Florence Chipman* passengers who were not from Wednesbury (a total of twelve families – mainly Irish who had been recruited in Birmingham but also including a Dutch civil engineer who, along with his wife and children, was recruited by Almeida Portugal in London) were sent to Cananéia, an agricultural colony that was soon to attract substantial numbers of English settlers.[12]

During the nineteenth century, the southern province of Santa Catarina was a favoured location for the establishment of both private and state sponsored agricultural colonies. Most of the province's inhabitants were concentrated on the coast, in particular in and around a few locations such as the island of Santa Catarina (the location of the provincial capital, Desterro), São Francisco do Sul and Laguna. The remainder of the province was even more sparsely populated, the only real centres of population being the small military garrisons that were located on what were considered strategically sensitive routes in the rugged highlands of the interior. Santa Catarina's population was ethnically diverse, its coastal inhabitants being predominantly the descendants of eighteenth-century immigrant arrivals from the Azores and of African slaves. Although Xokleng and other Indians lived in the province's densely forested interior, the authorities were always keen to dismiss the indigenous inhabitants as merely a primitive, scattered and nomadic people with no real ties to the land. The Xokleng's persistent acts of armed resistance in defence of the encroachment on their territory were usually dismissed as being nothing more than inconvenient and petty acts of banditry and no meaningful threat to what the Brazilian authorities regarded as the implantation of a permanent and civilized European population. What is certainly true, as John Hemming, an authority on the history of Brazilian Indians, has concluded, is that "the balance of terror was heavily in [the immigrants'] favour", with concerted efforts made by professional Indian-hunters to pursue and eliminate the Xokleng regardless as to whether they were an actual threat to settlement efforts.[13]

Lacking the plantation-agriculture tradition of tropical Brazil to the north, Santa Catarina's economy was a far more limited one, based on supporting the scattered military outposts, on semi-subsistence farming and fishing on the coast and on cattle raising in the interior highlands. During the course of the nineteenth century, the Brazilian government became increasingly keen on unlocking *terras devolutas* in Santa Catarina and boosting the province's population. To this effect, the government sought to encourage foreign immigrants to settle in agricultural colonies being carved out from legally unclaimed, state-owned land in Santa Catarina, a province located well-south of the tropics and thus ideally suited, so proponents held, for settlement by northern Europeans farmers.

The first such government-supported settlement was Colônia de São Pedro de Alcântara, established in 1828 with a mix of German mercenaries demobilized from the Brazilian army and their families, and immigrants newly arrived from Bremen. Located on the province's mainland a short distance due west of Desterro, the colony was placed along the road being constructed to connect Lages – in the heart of Santa Catarina's highland plateau, an important passage point for cattle herds being driven north from Rio Grande do Sul to São Paulo – with the coast. Other largely German agricultural colonies were formed soon after nearby, but the results were disappointing due to a combination of poor communications and insufficient financial support.[14] A similar fate met the privately-organized Belgian colony of Ilhota, founded in 1845 and attracting the first immigrants to settle in the Itajaí-Açu river valley. Although most of these immigrants abandoned the colony either for Desterro or the more dynamic colonies that soon emerged elsewhere in the valley, even thirty years later there remained at Ilhota some twenty Belgian families – described, by a French traveller, as being "nice Flemish figures, sunburnt and exhausted by hard work, but still recognizable".[15]

The 1850s and 1860s were again busy decades for the establishment of agricultural colonies in Santa Catarina, including several that would serve as examples of the wider potential for immigrant-based land colonization schemes. The privately organized Colônia Blumenau (founded in 1850) and Colônia Dona Francisca (founded in 1851, eventually to develop into the city of Joinville) were two of the most trumpeted examples of Brazilian land settlement achievements of the second half of the nineteenth century, again mainly attracting German immigrants. The social origins of the immigrants arriving in Blumenau and Dona

Francisca were much more diverse than had been the case with São Pedro de Alcântara and, in part due to on-going contacts with Germany, these two colonies ultimately developed into self-sustaining communities. But the problems that both colonies experienced, most notably substantial leakages of population to other parts of Brazil (especially northwards to Curitiba and elsewhere in Paraná), return migration to Germany, remigration to third countries, as well as a lack of profitability – an important issue if private-sector involvement in settlement attempts was to be maintained – were generally conveniently overlooked by proponents of immigration.[16]

One of the more successful of the government-managed agricultural colonies, at least in terms of population growth, was Colônia Brusque, in the northeast of Santa Catarina. Established as Colônia Itajaí in 1860 with 54 German *colonos*, the settlement was soon after renamed Colônia Brusque. Brusque's population steadily increased and by 1867 it had 1,333 inhabitants, overwhelmingly Germans (and their Brazilian-born children) from Baden and Schleswig-Holstein. Soon all land suitable for cultivation and settlement within the colony's existing boundaries had been allocated, and in 1866 additional *terras devolutas* adjacent to the colony were taken under the authority of the Ministry of Agriculture to form the basis of a new agricultural settlement, Colônia Príncipe Dom Pedro.[17]

Situated on the right side of the Itajaí-Mirim River, bordering Brusque, the centre of Príncipe Dom Pedro was located approximately 130 kilometres from the coast if travelling by river, or eighty kilometres taking the most direct overland route. The colony's territory encompassed a roughly triangular area, largely bounded by the Itajaí-Mirim and Aguas Claras rivers, with the village – consisting of merely a few houses, public and commercial buildings – located at the point where the two bodies of water joined. Close to the village the Itajaí-Mirim River was deep and some fifty metres wide, but the Aguas Claras was a powerful mountain stream, normally shallow but with a tendency to rise and broaden suddenly following heavy rains. It was almost immediately apparent that the new colony had insufficient publicly-owned land for distribution, most of that which was available consisting of the steeply mountainous and rocky valleys of the Serra de Bateas in the distant reaches of the Itajaí-Mirim River, areas entirely unfit for cultivation.[18]

Communications were poor, the colony's riverside location in many ways more of a hindrance than a benefit to development. The port of Itajaí, situated at least forty to sixty kilometres from the plots of land assigned to immigrants, could only be reached by boat or small barge when the Itajaí-Mirim's water level was low, and even then with utmost difficulty. When the river was swollen, it could not be navigated at all due to the ferocious force of the water. At the best of times, the journey was costly and slow, taking at least four days to complete a return trip. The so-called road – in reality barely a forest trail – that connected the colony to the mouth of the river, was only just passable and especially treacherous to mules, horses and their riders. Far from any real centres of population, there was no effective market and the cost of transportation alone made marketing produce unrealistic.[19]

Just as Príncipe Dom Pedro's location and the quality and quantity of land within the colony's limits were all wanting, so too was the administration. When possible, the Ministry of Agriculture appointed as directors of its state colonies men of the same nationality (or at least those who could speak the language) of the main body of settlers, the assumption being that they would understand each other's distinctive characteristics. The first director to preside over Príncipe Dom Pedro was Barziller Cottle who had been recruited in New York City along with his wife, Rebecca, his daughters and son, and had accompanied the first group of immigrant arrivals to the colony.[20] Isolated and far from higher administrative authority, a director of a state colony had substantial powers vested in him over such key matters as the allocation of plots of land to newly arrived immigrants, assigning employment in public works projects to men – crucial during the *colonos*' (pioneer settlers) early months while awaiting their first harvest – and the right to expel people deemed (by the director himself) to be causing a public nuisance. Like Príncipe Dom Pedro, other state colonies were all too often criticized for their poor locations, insufficient resources and a rapid turnover of administrative personnel, but few of their directors were quite as controversial as the self-styled "Dr" Cottle.[21] Said to be almost illiterate, Cottle's medical background was that of a surgeon, his experience gained on the battlefields with the Union forces in the recently concluded American Civil War.[22] Cottle was viewed with utter contempt, or even hatred, by many of the immigrants, who remembered him as "an unscrupulous ruffian, who demoralized his subordinates and

3. A "New Ireland" in Brazil

"Dr" Barziller Cottle in 1867.
(University of Nevada-Reno Library)

who systematically plundered the government and colonists, abusing, also, of the despotic powers of the directorship to oppress the settlers". Cottle was accused first by Príncipe Dom Pedro's settlers and ultimately by the government, of disregarding his duties, withholding rations, cheating in his accounts, borrowing (and never repaying) the savings brought to Brazil by the better-off immigrants, tampering with the *colonos*' correspondence (removing any letter that contained criticisms of his administration) and generally only ever pursuing one aim – that of self-enrichment.[23]

The allocation of work was a particular source of animosity, with Cottle being accused of embezzling considerable sums of Ministry of Agriculture money that should have been used to pay *colonos* engaged in public works projects. When Cottle did put men to work – such as on feeble efforts to up-grade the trail linking Príncipe Dom Pedro with Itajaí or on his personal holdings (or those of his son, David) – he was accused of favouritism in the selection of workers and even then withholding payments. The net result was considerable anger on the part of the *colonos*, adding to the frustration of the many completely idle immigrants who had expected on arrival in the colony to find land fully surveyed and ready for allocation.[24]

The first people to acquire land in Príncipe Dom Pedro were Germans from São Pedro de Alcântara and Itajaí, but the first immigrants to arrive directly from overseas were 98 Irish-Americans, as the New York recruits were often known, who came ashore in mid-February 1867 at the port of Itajaí together with Cottle and his family. With no preparations having been made for the group in Príncipe Dom Pedro – not even any basic tools being available – it was decided that the settlers should remain in Itajaí until the land was surveyed, and basic cabins erected. Although the members of the group had been selected in Rio de Janeiro as supposedly the best of the New York recruits, they nevertheless did not succeed in impressing the director of neighbouring Brusque, Baron Maximilian von Schneeburg, who immediately after their arrival commented that "the majority of the Irish *colonos* are people of the worst conduct, getting drunk, stealing from homes and fields, intemperate without limits, threatening with knives and pistols".[25] The president of Santa Catarina who simply could not understand why "vagabonds of every kind of moral defect" had been inflicted on his province shared this assessment of the newly arrived *colonos*.[26] Barroom brawls were commonplace and after an especially vicious rampage on 24 February, when drunken Irish immigrants attacked Germans and Brazilians, Cottle ordered the expulsion of sixteen Irishmen, charging them with "endangering the tranquillity and good procedure of the colony". While no doubt sparked by alcohol, this violent episode was a first outburst of protest at the situation in which the new immigrants had found themselves. The ethnic dimension to protests was to be repeated over the next two or three years in an atmosphere characterized by permanent suspicion, jealousies and tension that led to frequent minor acts of revolt.[27]

With so many frustrated, idle and angry men in the port of Itajaí, von Schneeburg decided to transfer the New York recruits to Príncipe Dom Pedro. A shed was hurriedly erected to provide temporary shelter and surveying teams organized. Cottle formed an armed militia of his most trusted men to protect the surveyors working in the far reaches of the

3. A "New Ireland" in Brazil

TABLES 3–5

When discussing English-speaking populations in Brazil, nationality is often a source of confusion. The term 'English' ('Inglés') was often synonymous with 'British'. As such, the Irish in particular are underrepresented in statistics. In the case of Príncipe Dom Pedro, further confusion surrounds the fact that virtually all the English-speaking immigrants were previously resident in either the United States or England. Given what is known about Príncipe Dom Pedro's immigrants, it is fair to assume that many of those listed as either 'English' or 'American' were born in the United States or in England of Irish parentage. Others may have chosen to identify themselves as 'English' or 'Americans', or Brazilian officials may simply have chosen these national identities for them.

TABLE 3: Nationality of immigrants arriving at Colônia Príncipe Dom Pedro, Jan–Dec 1867

Nationality	Number of arrivals	%
American	237	35.5
Irish	129	19.5
English	108	16.0
French	76	11.4
German	61	9.0
Italian	10	1.5
Swedish	8	1.0
Swiss	5	0.8
Canadian	5	0.8
Belgian	4	0.5
Others*	27	4.0
Total	670	100.0

* Includes Austrian, Danish, Norwegian, Scottish, Spanish.

Source: Lauth, *A Colônia Príncipe Dom Pedro*, p. 98.

TABLE 4: Nationality of immigrants arriving at Colônia Príncipe Dom Pedro, Feb–Oct 1868

Nationality	Number of arrivals	%
English	249	55.4
Irish	114	25.3
German	37	8.2
American	22	4.9
Dutch	7	1.6
French	6	1.3
Others*	15	3.3
Total	450	100.0

* Includes Austrian, Canadian, Italian, Swedish and Swiss.

Source: Lauth, *A Colônia Príncipe Dom Pedro*, p. 98.

TABLE 5: Nationality of immigrants arriving at Colônia Príncipe Dom Pedro, Jan–April 1869

Nationality	Number of arrivals	%
English	13	61.9
Scottish	5	23.8
Irish	3	14.3
Total	21	100.0

Source: Lauth, *A Colônia Príncipe Dom Pedro*, p. 98.

colony from raiding parties of Xokleng Indians and also to hunt deer and wild pigs to provide the settlers with supplies of fresh meat. On 9 March 1867, less than a month after the first group of immigrants from New York arrived in Santa Catarina, Príncipe Dom Pedro received a further nineteen immigrants from New York to replace those who had been expelled, increasing the Irish-American contingent to 101 people.[28]

Most of the immigrants from New York were single men, mainly unskilled labourers or mechanics. Their apparent lack of agricultural experience was, however, of little immediate importance: when Príncipe Dom Pedro was visited in March by Ignâcio da Cunha Galvão, the Ministry of Agriculture's inspector of colonies, he found the immigrants still housed in the reception shed, awaiting the allocation of land. The encampment was located in the centre of the colony, alongside the Aguas Claras River, in the pasture of the sawmill owned by a long-time resident, Franz Sallentein. With a lack of publicly-owned land suitable for cultivation, Galvão recommended that the government purchase Sallentein's property, the best in the colony, which had only really been used for the extraction of timber. After a wait of almost three months, the Ministry of Agriculture agreed to this proposal, but the Sallentein property was only sufficient for the creation of ten or twelve allotments of the usual twenty to thirty hectares per family – far too few units to satisfy the needs of all amongst the first party of immigrants, let alone provide for the 670 people sent to Príncipe Dom Pedro during its first year.[29] Despite a high turnover of immigrants arriving in the colony during its first few months, some settlers did persevere. A correspondent of the *Kolonie Zeitung*, a German-language newspaper published in nearby Joinville, was surprised to find himself acknowledging that, although some of the Irish-American *colonos* had behaved as "drunken thugs", those who remained once the trouble-makers had been disciplined or expelled were generally good, hard-working, people who were already showing impressive signs of adapting well to Brazilian conditions and demonstrating that "they have more practical sense than our good German peasants".[30]

Word soon reached Desterro of the Irish presence in Príncipe Dom Pedro, news that attracted the interest of Joseph Lazenby, an Irish Jesuit priest at the Colégio do Santissimo Salvador. Aware that Príncipe Dom Pedro was without a chaplain, at the end of the year Lazenby made his way to the colony where he set to work improving, as he saw the task, the spiritual well-being of the immigrants whom he considered had been overcome by "the brutalizing British and Irish vice of drunkardness [sic]". Travelling by horseback around Príncipe Dom Pedro, within a period of a mere fortnight Lazenby claimed some remarkable success in the raising of morale, declaring that "by preaching, confessional, and private exhortations I have put down drunkardness [sic] to a great extent". By the end of the year – a period of just a very few months – Lazenby claimed to have successfully established within the colony a temperance society, opened both a primary school and a church on a hillside overlooking the village and had started work on building a curate's house beside the colony's cemetery. Perhaps most remarkably of all, in mid-February 1868 Lazenby baptized into the Catholic church the Cottle family – an act that the fiercely anti-clerical Prussian diplomat Johann Jacob Sturz suggested was merely a cynical move by "Dr" Cottle to ingratiate himself with his government superiors. But whether his true motivations were practical or spiritual, "Dr" Cottle's conversion came too late to influence his role as administrator, although his new-found religious zeal led him to call for the Príncipe Dom Pedro to be officially declared a Catholic colony, a call that the authorities in Rio de Janeiro rejected.[31]

As Lazenby's services were required back in Desterro, he realized that a permanent priest must be found for Príncipe Dom Pedro. It was at about this time that Lazenby entered into contact with George Montgomery, encouraging him to take up the vacant position of chaplain and to direct to the colony emigrants from Wednesbury. Lazenby held out high hopes for any people Montgomery might send out, believing that they would "help to give a tone to the colonists [from New York], many of whom got perverted in their conduct whilst in the United States".

Although Lazenby admitted that some of the *colonos* had experienced trials and unfortunate upsets, he thought most of these to be of their own making, believing that they could be overcome with support and direction of the sort he had provided during his period in the colony and which hopefully Montgomery would provide in the future.[32]

Despite Príncipe Dom Pedro's difficulties, a small riverside village gradually emerged. In addition to the school, church buildings and cemetery, Príncipe Dom Pedro's incipient urban centre boasted by late 1867 administrative buildings, two large stores, a "hotel" (albeit one that offered comforts not much better than a shed), a butcher's shop, a bakery, a blacksmith's forge, a gristmill, a sawmill, several houses and, especially appreciated, a tavern. Rough trails were beginning to open up the colony's hinterland, forest was gradually being cleared and land planted. Houses were replacing the huts on the allotments and some of the *colonos* boasted a few pigs, some poultry and perhaps even a horse. Tools were made locally for working on the land and for sale to *colonos* elsewhere.[33] Recognizing the need for the immigrants to be able to communicate for trade or negotiations beyond the narrow confines of the colony, classes were organized to teach the adults Portuguese.[34]

At precisely the same time as Montgomery's Wednesbury party was leaving Birkenhead for Brazil, other immigrants – fifteen English families (a total of around forty people) – were being dispatched from Rio de Janeiro to Príncipe Dom Pedro. Arriving in the colony in February 1868, the families faced the same fundamental obstacle that the Irish-Americans had experienced – that the only lands still available for distribution were the swampy bottoms and steep mountainsides of ravine-like valleys, a fact that Lazenby was withholding from Montgomery. A few of the new arrivals purchased allotments from settlers beginning to drift away from the colony, but most accepted whatever surveyed portions of land were available, regardless of their agricultural potential. Compared to both the New York recruits and the Wednesbury Irish who were to arrive later, these families were relatively affluent. Most had paid their own passage from Liverpool to Rio, apparently attracted to Brazil by the emigration agents Meadows and Christopher or by advertisements in *The Times* of London placed by Joaquim de Almeida Portugal. Calling for "men of good character", these notices had explained that "capitalists can purchase at very cheap rates tracts of superior land suitable for agricultural and pastoral purposes, and passages secured at low rates" and in all likelihood appealed to families unable to afford to settle in the United States or in a British colony in a style that they would have regarded as appropriate to their status. With five of the families carrying between them savings of some £1200, the group attracted Cottle's particular interest and he somehow succeeded in persuading them to lend him most of their capital. True to form, the loan was never repaid. Fearing retaliation from the immigrants who felt duped, cheated or otherwise abused by the colony's autocratic director and also now facing the possibility of investigation by the authorities in Rio, Cottle fled with his wife to the United States, eventually finding refuge in Eureka – a remote silver-mining community in central Nevada.[35] While the charges directed at Cottle were serious, animosity was certainly fanned by rivals in the colony's administration. One particularly strong critic was Elpídio de Mello, the then engineer of Itajaí, and a future director of Colônia Príncipe Dom Pedro. After having been publicly labelled a thief by de Mello, Cottle challenged his accuser to a duel, in turn accusing him of spreading false rumours and fomenting intrigue within the colony, organizing secret meetings and petitions and displaying ridicule and contempt towards the Catholic religion. There is no record as to whether the challenge was accepted or acted on.[36]

Within just a few months of their arrival in Príncipe Dom Pedro, most of the English families were preparing to leave for Argentina or Uruguay. Without any savings or, in most cases, personal belongings to sell, the Irish-Americans were also seeking ways of escape. Following the flight of Cottle in early March 1868, Galvão returned to Príncipe Dom Pedro where he succeeded in persuading some of the English to give the colony a second chance. But it was not long before they reversed their decision. Crops failed and the *colonos* continued to find it impossible to obtain from the government payments owed to them for employment on public works. At times there was not any food available in the colony, even for the few *colonos* with the means to purchase it, and so hungry were they that many were reduced to a diet of palmetto (hearts of palm), weeds and bark.[37]

Despite Scully's warnings regarding the suitability of Santa Catarina, but presumably just before reliable reports of the worsening mood and conditions in Príncipe Dom Pedro had filtered out, the *Florence Chipman* immigrants arrived at the port of Itajaí on 3 May 1868. There they met some of the departing English settlers who warned them against proceeding further. Perhaps not wanting to believe what they were being told, and certainly without the financial means to return to Rio, let alone England, even had they wanted to, the members of the Wednesbury party allowed themselves to be escorted to the colony – the "New Ireland" for which they had been told in England they would be the vanguard.[38]

Just as the Wednesbury group arrived in Príncipe Dom Pedro, conditions in the colony were becoming ever more hopeless – in July 1868 the Brazilian Liberals lost power to the Conservatives, leading to greater hardships in Príncipe Dom Pedro. In reality, the two political parties possessed almost identical policies and programs, not least in terms of their support for the institution of slavery and their erratic attitudes towards the promotion of immigration and the assistance to be provided to immigrants on arrival. But a new government in Rio de Janeiro inevitably led, in a politically highly centralized Brazil, to the change of virtually all personnel at every level of public administration. This led to a major disruption of services. The new minister of agriculture, Joaquim Antão Leão, was determined to make large-scale budgetary cuts in the support of state colonies, in part due to the mounting costs of the Paraguayan War, but also due to his opposition to assisting immigrants of no obvious benefit to plantation owners. In September, Leão halved the funding of public works in the state colonies, and in the following April it was suspended altogether, cancelling the claims of *colonos* for work owed and for the cash bonus that had been promised on taking up lots and erecting houses. In Príncipe Dom Pedro, the government store which had offered *colonos* credit for the supply of provisions (debts to be repaid once land came into production) was closed with the effect, according to John Haher, the colony's newly installed chaplain, of driving away yet more *colonos* as increasing poverty left people on the brink of starvation.[39]

Recognizing the consequences that the withdrawal of funding would have on the already struggling colony, *The Anglo-Brazilian Times* was uncharacteristically critical of the Ministry of Agriculture, describing the decision as one of "heartless cruelty".[40] In Príncipe Dom Pedro itself, Haher was at a complete loss to understand why the Ministry was financially abandoning the colony after, "enticing poor people to come to [Brazil] promising them advantages of all kinds, for years, or for such time till they could live independently by their labour".[41]

As conditions deteriorated yet further, jealousies and factionalism within the colony increased. The interim directorship of the colony had passed briefly to Baron Frederick von Klitzing, the former stable master of Maximillian, archduke of Austria and tragic emperor of Mexico. After Maximillian's fall from power and execution before a firing squad, von Klitzing escaped to Brazil. There a position as director of Brusque was found for him, perhaps as a result of pity for a fellow imperial courtier. To some of the Germans at least, von Klitzing – described as "an energetic, well disposed man" – was just what the colony needed. There was a feeling that there could not have been a greater contrast than between the baron, a "genuine German aristocrat [...] who has a sense of law and order" and Cottle, a "Yankee speculator [who], thanks be to God, we got rid of".[42] Elpídio de Mello, who took over as director of Príncipe Dom Pedro in May 1868, sought to maintain von Klitzing's firm style of leadership. People he found congregating in the village's praça without any real reason (usually drinking) were moved on and several persistent "trouble-makers" were expelled altogether from the colony. But although he was supported by some of the *colonos*, de Mello's strength was sapped by constant demands for payment for public works that he was unable to meet due to continued lack of funds and an absence of support from officials both locally in Desterro and in Rio de Janeiro.[43]

After de Mello's departure in September 1868, responsibility for the management of Príncipe Dom Pedro was again given to von Klitzing, who was temporarily relieved by the appointment of yet another director, Manoel Moreira da Silva.[44] Von Klitzing, also the subject of suspicions and hostility, was accused of unconcealed favouritism, in particular of providing what very few employment opportunities were still available on Príncipe Dom Pedro public works to Germans from Brusque, rather than enlisting locals. But even when residents of the colony were given jobs there was internal ill-feeling with, for example, the New York Irish said to be favoured over the Irish from England, while the Germans were again accused of being the most privileged of all. Certainly the colony was in a desperate condition, the Irish and English *colonos* at least united in feeling frustrated by an inability to express their grievances to the colony's administration – none of whose officers spoke any English – and the apparent lack of interest in their plight shown by the settlement's director.[45] Allegations of corruption continued, with some *colonos* complaining that land they had cleared and started to cultivate had been appropriated at the whim of the colony's director and replaced by hillside forested tracts.[46] The *colonos* regularly gathered together to voice their frustration, the demonstrations attracting a hundred or more angry people. Unsympathetic to the troublemakers, as he considered them to be, von Klitzing, complained in a report of 27 March 1869 of being "forced to go every day, sometimes twice a day, and even in the night to break-up illegal gatherings". In the middle of the following month, von Klitzing tried to dampen protest by expelling the "sluggards of the colony" whom he felt were stirring up trouble.[47] While there were no doubt some ringleaders amongst the disgruntled immigrants, it was unfair to reduce this disaffection to the nuisance caused merely by a few troublesome individuals.[48]

In early 1869 *colonos* from Príncipe Dom Pedro began appearing at the office of Charles Watson, the British vice-consul in Desterro and at Lazenby's college, to appeal for help. For some of the immigrants at least, the only hope of salvation for Príncipe Dom Pedro was to declare it a Catholic colony – as Montgomery had expected and as Cottle claimed to have desired – and to cede responsibility from the government to Lazenby and his colleagues. In July the *colonos* remaining in Príncipe Dom Pedro sent a deputation of fifty men to the provincial capital to explain to the authorities that the terms which they had agreed to had been totally ignored, to demand payment from the government for work that they had completed and to ask for the reintroduction of the public works' schemes. The provincial president, however, refused to meet with the deputation, offended by their boisterousness and making of a public protest that breached basic protocols of the hierarchically obsessed Brazilian society. Instead the group appealed to Watson to make representations on their behalf. With the protesters hungry and without shelter, Watson agreed to pay personally for their food and lodging. At first the president maintained his stance of refusing to meet with the demonstrators and completely dismissed claims for payments dating from Cottle's corrupt administration. A compromise was finally reached whereby the *colonos* agreed to return to Príncipe Dom Pedro in exchange for a guarantee that work would be available with payment in the form of supplies sufficient to tide them over until the next harvest. But within a matter of months any remaining hope for the colony had evaporated. The promised provisions were not delivered and the *colonos* refused to do any more work, certain that they would never be paid.[49]

While some of the immigrants from New York and Wednesbury remained in Santa Catarina by reason of choice or lack of any other possibility, others sold their few personal belongings to return to Rio in the hope of somehow securing a passage to England or, preferably, to the United States. Arriving individually or in groups consisting of as many as 25 people, utterly destitute – and sometimes quite literally starving – Irish families were a common sight in the streets of Rio as only a few of the former *colonos* were allowed to return to the Casa de Saúde and then under the strict condition that they would not be issued with rations. All of the returnees had experienced enormous suffering, with most having lost a husband or wife, a father, a mother or children to hunger, yellow fever, smallpox or other illnesses while in Príncipe Dom Pedro.[50]

British diplomats claimed to be at a complete loss as to how to help. The Foreign Office had always taken a rigid stance not to support those whom it regarded as derelicts abroad and so, with consuls having no official financial resources to draw upon to assist distressed British subjects, they had to appeal to wealthy members of the British community and private charities – resources that were now overwhelmed by this sudden demand, their being in normal times focused on helping stranded British sailors. But the state of the former *colonos* shocked, or more likely embarrassed, Rio's overwhelmingly prosperous British community, unused to seeing compatriots "starving" in the streets, "many of them without even rags to cover them". A public subscription was started for their relief, with sums of between £5 and £10 donated by British individuals and companies, with the British minister, George Buckley Mathew, heading a list of contributors which included representatives of leading British merchant houses and banks in Rio.[51] This was the start of a period lasting several years when destitute Irish and English immigrants would be a constant headache for the established British community in Brazil. Just as Charles Watson was supporting former *colonos* in Desterro from his own pocket, so the British consul in Santos, where other immigrants had fled, responded similarly. So financially extended was he that he was eventually reduced to taking in a gentleman lodger to supplement his income.[52] Mathew in Rio was later to recall "the state of our duped emigrants whose misery has ruptured my pockets".[53] Most of the Irish recruited in New York demanded to be sent back to the United States, leading one former Confederate to remark with a mix of contempt and wry satisfaction:

> The [ones] that returned were the scum of New York, drank up all their money in whiskey and then swore the Govm't [sic] had fooled them and asked Mr Yankee consul to ship them back to the best country the world ever saw. All I say [is that] I feel pity for you all – to have a miserable set of humans let loose on you. But America is now the hell of earth and all good devils go there.[54]

Meanwhile in Príncipe Dom Pedro, the remaining *colonos* concentrated on sheer survival. In desperation, a few somehow maintained feelings of hope for the colony's future, having purchased or laid claim to one of the better of the abandoned lots. Unable to market their crops, they could sometimes barter their meagre surpluses of maize, beans or manioc for cachaça (sugar cane spirit), clothing or other basic requirements.[55] A few generated a small income, having established, taken over or being employed in one of the few businesses in the colony. But most of those who had stayed on had neither the funds nor the energy to leave and were resigned to whatever the future might bring.[56]

As if neglect of Príncipe Dom Pedro by the central and provincial governments, maladministration, unsuitable immigrants, lack of markets and insufficient land were not enough, the location of the colony meant that those *colonos* who persevered were subject to the whims of nature. All the settlements in the Itajaí Valley were prone to flash floods, but Príncipe Dom Pedro's situation between the Itajaí-Mirim and Aguas Claras rivers made it especially likely to meet disaster. In the middle of the night on 27 November 1868 the inhabitants of both Príncipe Dom Pedro and Brusque were suddenly awoken by torrents of water raging down into the valley from what were normally gentle mountain streams. For the past five months it had rained virtually every day and the ground had become saturated. Without warning bridges and livestock were swept away and the roads and trails became impassable. One of the German residents of Príncipe Dom Pedro lamented that it was the "decent, hard working" *colonos* who suffered most, and explained that "the first harvest that the good people of the colony had hoped for is now doubly awful" whereas the "rabble" had nothing to lose anyway. One immediate consequence was an increase in the already high prices of food and other necessities. Unwilling either to assist or evacuate the *colonos*, the Ministry of Agriculture did little – apart from channelling funds for the installation of oil lamps along the streets of the town's small centre. Given the deteriorating state of the community, this curious priority led a German *colono* to comment how "the glaring street lights only serve to illuminate the dreadful state of the colony's economy".[57]

As immigrants continued to drift away, any lingering hopes for Príncipe Dom Pedro vanished over the course of a twenty-four hour period in 1869 when, once again, flash floods struck the colony, a devastating event that was recorded by Peter Hayes:

> On the last Thursday of the month of November, about the hour of twelve (noon) the sky (which had been very clear) suddenly darkened and it began to rain. Rain in such torrents as I never before witnessed in my life; accompanied by loud peals of thunder, all of which continued at intervals for twelve hours. The river Itajaí began to rise with fearful rapidity until it reached the village....[58]

The flood-waters wreaked destruction and despair on an even greater scale than had occurred the year before. All the trails leading to the outlying homesteads were cut, preventing any possibility of providing warning or assistance that people in the village might otherwise have been able to offer the more isolated *colonos*. As before, people were woken in the middle of the night by torrents of water. The village buildings were quickly abandoned as the river's water level almost instantly rose to three metres in depth, flooding the land around. With no time to gather up their belongings, their occupants fled to the few nearby houses, located on higher land, that were not completely submerged. The hostel, although situated on rising ground, was inundated by flood-water to the level of its reception desk, while the house belonging to Dr Waidell, the colony's surgeon, had water over two metres in depth. Attempting to escape, the raft onto which Dr Waidell had clambered overturned and he was forced to swim some eighty metres to safety, just managing to save his life. Somehow he made it back to his house, climbing onto its roof. The sawmill was completely destroyed and the gristmill severely damaged, with their dams completely swept away along with all equipment and their stores of timber, corn, flour and sugar. The slaughterhouse too was smashed to pieces by the force of the water and the government administrative offices were submerged. Similarly, the outlying homesteads were severely affected. The *colonos* along the Aguas Claras River sustained considerable losses: woken by the torrents of water without any warning, many people fled their homes with only the night-clothes that they were wearing; their houses, poultry, pigs and what little money, clothing and other personal possessions they still had, all taken by the flood waters. The trail leading to the Aguas Claras homesteads was washed out, making travel impossible and leaving the *colonos* completely isolated. But it was the inhabitants alongside the Cedro River who suffered the most, losing property, crops and, in some cases, their lives. The people here were awakened at midnight as water entered, within moments rising to the level of their beds. Unable to reach higher ground, the *colonos* climbed onto their houses' rafters and then to the roofs, to escape the rising water. Some of the homes were swept away, including that of the Hopkins family, where the Kirby family had also vainly taken refuge. Although Mr Hopkins and Mr Kirby managed to save themselves by grabbing onto a floating tree, Mrs Hopkins and two children were swept away and drowned, their bodies only found days later when the floodwaters receded. Another *colono* who died was John Haskell who, with a neighbour, was in the floodwaters for sixteen hours before he was able to reach land. Taken to the house of Thomas McDermott, Haskell died of exhaustion half-an-hour later. Losing all their few possessions, the colonos were utterly destitute.[59]

With most of their homes and fields wiped out, the survivors of the flood were faced with a stark choice: remain in Príncipe Dom Pedro and rebuild on the land by rivers that they knew could be deadly, or leave. In one final act of protest, the departing immigrants drafted a petition, reiterating their complaints concerning payments owed to them and criticizing the current director for his "bad and ungentlemanly management". He had been able to find money "to pay the many useless officers while the poor Colonist is made to suffer", and feel frustrated for "not being able to get civel answers to Civel Questions" [*sic*].[60]

Exhausted as the immigrants were by their experiences in Príncipe Dom Pedro, Rio de Janeiro was not always the safe-haven they must have prayed it would be. Instead, conditions in Brazilian capital often delivered yet more suffering. Wednesbury emigrant Martha Connolly Orgill was one of those who became dangerously ill while making arrangements to leave Brazil, but in a poignant letter to her sister in

3. A "New Ireland" in Brazil

Liverpool she explained that it was her two children who suffered far worse a fate:

> We had arangements to start from Rio in an English Steamer to New York on the 19 of January and had eavery thing ready only waiting for the time but the children had had the hooping cough about two weeks then and on the 13 they Broak out with the small pok. Poor little Mary commencen with convulshouns and was neaver out of them untill she died. She did not know eather of us all the wile. Their was no hope of Mary from the first but great ones of Joseph but poor little pet he died the first on the eavening of the 19th of January and our prettey little Mary on the morning of the 20th. They were both burred togeather in the same grave. Mary was 3 years 2 months Joseph 1 year and 4 months and My Dearest sisters I sincerley hope you will neaver have the trouble to goe through I havd to sit and watch my two prettey ones deaying together and did not know which to pay the most attention to or which would leave me first. I was that exhausted with sitting up for 7 days and 7 nights for you must know while life was in them I could not leave them. I had to lye down at last bye My Dear Marys side and close her sweet little eyes in Death. I can asure you it has been a very heavey blow to us both for in all our troubles she was a great comfort. She was so old fashioned she was just like an old woman. It was very hard after bringing them through all the trouble and to loos them at last.[61]

Soon after the tragedy of losing her children, Martha, with her husband Austin, followed others from Príncipe Dom Pedro to the United States, settling in western Pennsylvania.[62]

With news of the floods and survivors reaching the capital, the British minister pleaded with the Brazilian government to evacuate Príncipe Dom Pedro, return the remaining *colonos* to Rio and provide them with passages back to England, to the United States or elsewhere abroad.[63] Alarmed by reports of sickness and hunger, the British vice consul in Santos, Elliot Bushby, was sent to Santa Catarina to investigate. When Bushby's steamer arrived in Itajaí on 20 May 1870, word of his mission rapidly spread amongst the Príncipe Dom Pedro *colonos*. Within days, 27 of the New York Irish and 85 English and Irish (who ranged in age from babies of just a few months to men in their fifties) boarded the vessel to accompany Bushby to Desterro, leaping at this opportunity to escape the colony for good. A further eight English and Irish families – a total of forty people – opted to remain in Príncipe Dom Pedro with the intention of harvesting a final crop, though a few of these stayed still longer. On arrival in Desterro, the provincial government housed the refugees in a large shed while a decision was being made as to what to do with them.[64]

On 1 June, Santa Catarina's president issued Bushby with an ultimatum. If the immigrants in Desterro continued to refuse to return to Príncipe Dom Pedro or were not prepared to accept transfer to an alternative state colony, then all government provisions would be halted. Bushby objected, demanding that the immigrants be transported to Rio, a reasonable request, he felt, in view of the government's breaches of contract. Meanwhile, a telegram appealing urgently for supplies reached Bushby from *colonos* camped out at Itajaí saying that they were now completely without food. The *colonos* also directed a petition to the British consul in Rio, desperately requesting medical supplies and food or, "God only knows what the consequence will be". With the provincial president refusing to alter his position, Bushby took responsibility for feeding the immigrants who had accompanied him to Desterro.[65] On the 27 June, Bushby arrived back in Santos along with 83 of the former Príncipe Dom Pedro *colonos*. The president of São Paulo agreed to provide the immigrants with food and shelter, while Bushby organized employment. Most of the men were given work on the British-owned Jundaí to Campinas railway, while the unmarried women found employment as domestic servants by British or American families. Of those who chose to remain in Santos, several men were recorded as finding work as stevedores, while the children over the age of ten found work as domestic servants. Others decided to return to Rio, working their passage on board a coastal streamer, while two families opted to accept offers of re-settlement in Cananéia where a few of the *Florence Chipman* party had been settled.[66] His duty done, Bushby claimed to be optimistic as to the immigrants' prospects, "provided they work and abstain from drinking the cheap but intoxicating rum of the country".[67]

Eventually all of the Irish *colonos* remaining in Santa Catarina left the province. Many made their way to Rio from where some were evacuated, or worked their passage, to the United States.[68] Of these, most settled in western Pennsylvania where they worked in the emerging steel industry, the Wednesbury contingent returning to occupations similar to those that they thought they had left behind in England. A few found work in the port of Santos or nearby in the small, but growing, city of São Paulo: "Their poverty and misery is painful to witness," reported a fellow British resident, "two little boys appeared at the school with bare feet, the rest of the children being excited and surprised".[69] Others moved elsewhere in South America, joining the sizeable and well-organized Irish and English communities in Argentina or accepting the Ministry of Agriculture's offers of relocation to other state agricultural colonies in Brazil. Of these most were sent north to Assunguy, in neighbouring Paraná, meeting up along the way with some of the hundreds of other immigrants (whose circumstances are discussed in the following chapters) being sent there directly from England.[70]

Meanwhile in England Montgomery's emigrants effectively disappeared from the public eye. *The Universal News*, so keen on the idea of a 'New Ireland' in South America, chose not to report on the fate of the Wednesbury party, while Montgomery, his health fast failing, set aside his intentions of travelling to Brazil and remained conspicuously silent on the turn of events.[71] Nor were other voices to be heard reporting on the fate of the Irish in Santa Catarina, enabling similar schemes to be hatched in England with dire consequences for many of their participants.

PART IV

These white slaves of England with the darkness all about them, like the children of Israel waiting for someone to lead them out of the land of Egypt.

Joseph Arch
Leader of the National
Agricultural Labourers' Union
Wellesbourne, Warwickshire
14 February 1872

We are simply 'white slaves' – we work our lands for Brazilian task masters and leave our children for their heritage ignorance.

Frederick Tertius Tigar
English immigrant
Assunguy, Paraná
13 July 1875
[NA/PRO FO128/108/292]

Chapter 4
Agricultural Labourers in Mid-Victorian England

REGARDLESS OF THEIR official status and precise relationship with the Brazilian government, Brazilian emigration agents acting in England, as elsewhere, were adept at identifying potential recruitment grounds. In 1872 and 1873 Brazilian agents were again active in England, this time in the picturesque villages and small towns in the south of the country. It made perfect sense to supply Brazilian agricultural colonies with new immigrants from among the region's increasingly discontented and surplus rural and semi-rural labour force. Following the disastrous experience in Brazil of the Black Country emigrants, dissatisfied English farm labourers appeared to be altogether more suitable, and Brazilian agents sought to take advantage of the rural discontent. They found willing ears. As George Buckley Mathew, the British minister to Brazil, was to remark of the striking labourers, "after a constant attempt [of demanding wage increases] for some weeks [men] had nothing left before them but the workhouse – or Brazil!"[1]

Rural England in the nineteenth century presented a distinctive three-tiered social arrangement of landlord, farmer and landless labourer.[2] This social pattern had been developed over the course of centuries, but became more pronounced during the eighteenth century, in part from the decline of the yeoman farmer – the occupier of a smallholding – who had formed an important intermediate class, in part from the extension of enclosures to open fields and commons, removing access to property. In the course of a few generations, the English rural population was transformed into a demoralized populace, as explained by historians Eric Hobsbawm and George Rudé:

It is difficult to find words for the degradation which the coming of industrial society brought to the English country labourer; the men who had been 'a bold peasantry, a country's pride', the sturdy and energetic 'peasantry' whom 18th century writers had so readily contrasted with the starveling Frenchmen, were to be described by a visiting American in the 1840s as 'servile, broken-spirited and severely straitened in their means of living'. [....] From that day to this those who observed him, or who have studied his fate, have searched for words eloquent enough to do justice to his oppression.[3]

By the nineteenth century, relatively few farmers in England owned their own land. Instead, most rented from large landowners and employed a variety of skilled and unskilled workers, who made up the great majority of the rural population. At this time intensive agricultural modernization took the form of an acceleration of enclosure schemes – the appropriation of what had generally been considered common land – and the cultivation of larger units utilizing new methods of crop rotation and machinery. A significant consequence was the creation of a large surplus of rural labourers, a surplus that even the rapidly expanding British industrial centres were unable fully to absorb. Whereas at the beginning of the century a labourer might reasonably expect to plough the same fields or tend the herds of cattle of the same owner for years on end – perhaps even for an entire working life – by the mid-century labourers were becoming increasingly nomadic, selling their increasingly unskilled services to farmers for a day or at most a season at a time, finding themselves, in the words of a contemporary journalist, "ill-paid, ill-fed, ill-housed, and over-worked".[4]

Across England, considerable regional variations existed in rural social structure, pay and conditions, with small holdings continuing to be important in parts of the north, and with higher than average wages offered in counties such as Lancashire and Yorkshire, where alternative employment in industry lured away farm labourers. Even on a county level there were considerable wage variations with, for example, labourers' pay in Gloucestershire villages close to the city of Bristol being higher than those in areas further away. But overall, between 1867 and 1873 the average English agricultural wage, adjusted for differences in retail prices, was just 52 per cent that of the average industrial rate, while without these adjustments it was a mere 47 per cent.[5]

4. Agricultural Labourers in Mid-Victorian England

The working and living conditions endured by agricultural labourers in southern and central England, where alternative employment possibilities were extremely limited, were almost always harsh. An American traveller in the 1860s described the conditions endured by a Wiltshire farm labourer:

> His ordinary wages were eight shillings a week, and an additional shilling in haying time. By milking cows as an extra job, he was earning about eleven shillings, working from sun to sun. His wife received eight pence per day, making up fifteen shillings per week. He paid 1s 6d. weekly for house rent; had a garden-patch on which he had grown from twenty to twenty-five bushels of potatoes. His food was entirely bread and cheese on week-days for breakfast, dinner and supper. On Sunday he had a piece of bacon for dinner. He never spent anything for beer at the public-house, and drank only what his master allowed him, which was three half-pints a day. He could not lay by any of his wages for old age or sickness, do the best he might. When he came to be too old to work, he should depend upon the parish for out-door relief [....] His head was bent towards the ground upon his load as he thus spoke of his earthly lot and expectations. He had fought all the forenoon battles of life, and spoke of the last tug of the war with want and poverty with a cheerful voice, uttering no complaint nor a word against landlord or parish authorities. Indeed he seemed to have realized all the enjoyment he expected when he looked forward from the threshold of the life to which he was born. And who could have the heart to break the even spell or darken the dreams of this content in such a man's heart![6]

Although wages, for those labourers fortunate enough to have longer terms of employment, might be supplemented by such benefits as an allowance of milk and food at a reduced rate, a potato patch or vegetable garden, these benefits were not always easily managed and could place labourers into a position of indebtedness. Labourers who had allotments could usually only tend them during part of Sunday (though working on the Sabbath, even in one's own garden, was often frowned upon), and at meal hours in the week. Some farm labourers were able to keep a pig or two, but more often than not the animals were raised for sale to pay off debts rather than used as food for their families. Hardy's Tess worked on her father's allotment growing potatoes:

The plot of ground was in a high, dry, open enclosure, where there were forty of fifty such pieces, and where labour was at its briskest when the hired labour of the day ended. Digging began usually at six o'clock, and extended indefinitely into the dusk or moonlight.[7]

Such allotments were often located some considerable distance from a labourer's home and many hours could be spent simply walking to attend to the plot. During the harvest season when all farm hands were required from dawn to dusk, labourers could give their allotments even less attention than normal despite this also being the crucial period for their own crops.[8]

While there were regional variations, farm labourers were notoriously conservative in their food habits with diets that were almost always exceedingly monotonous. Among the southern English there was a heavy reliance on bread – bread and cheese, bread and milk, bread and bacon, bread and dripping. Fuel was often scarce and expensive and as workers could not as a matter of routine afford fires, there was less hot food. Potatoes would have been an obvious staple but were unsuitable because they required cooking (in Ireland peat was usually readily available). By mid-century, home baking had become increasingly rare with bread being supplied by the newly-appearing village shop or bakers' carts. Milk, unless supplied free by the farmers, was consumed less than might be expected and during the early decades of the nineteenth century tea became the most common drink, at least of the women and children. In extreme poverty burnt toast was soaked in hot water to produce a drink that vaguely resembled it.[9]

To many workers the daily allowance of alcohol – usually beer, or in the West Country apple cider – was an important consideration in accepting employment, with the quantity being as much as half a gallon or more a day. At harvest time a barrel from which the workers could help themselves was often placed in the corner of a field. The drink was not only something to look forward to (in part, perhaps, due to its addictive qualities) but helped to wash down the tedious midday meal of bread and cheese and provided valuable nourishment. Temperance reformers – including many union leaders – railed against the custom, but drink was a traditional and much appreciated part of the wage and it could be difficult to get workers to start a job without it.[10] Visiting England in 1850, an American traveller tried to persuade a labourer that

4. Agricultural Labourers in Mid-Victorian England

Field hands taking a refreshment break.

he and his family would be far better off in the United States. The writer's arguments, however, were rejected once it was discovered that American farmers did not give their employees a free allowance of alcohol:

'And how much beer?'

'None at all!'

'None at all? Ha, ha! He'd not go then – you'd not catch him workin' withouten his drink. No, no! a man 'ould die off soon that gait.'

It was in vain that we offered fresh meat as an offset to the beer. There was 'strength', he admitted, in beef, but it was wholly incredible that a man could work on it. A working-man must have *zider* or beer – there was no use to argue against that. That 'Jesus Christ came into the world to save sinners,' and that 'work without beer is death,' was the alpha of his faith.[11]

While at first sight the villages may have appeared picturesque, their sanitary conditions were almost always appalling, with open sewers running into the sources of drinking water for livestock and villagers alike.[12] The shelter provided workers was as often as not appalling, even to the eyes of contemporary observers. In 1872, a visit to a typical home of a Somerset agricultural labourer provoked the following description:

> The total length of the miserable hut was about seven yards, and its height, measured to the extreme point of the thatched roof, about ten feet; the height of the walls, however, not being so much as six feet. From the top of the walls was carried up to a point the thatched roof, there being no transverse beams or planks. In fact, had there been any, I could not have stood upright in this hovel. There was, of course, no second floor to the place, and the one tiny floor was divided in the middle into two compartments, each about three yards square; one used for a bedroom and the other for a sitting-room.[13]

Half-starved striking labourers, evicted from their tied cottages – accommodation that in reality could be no better than a hovel – contended with "an almost indescribable feeling of bitterness against them on the part of the squirearchy, the clergy, and the farming class [who] received the cordial support of the magistracy".[14] Unable to accumulate savings and dependent on their employers for shelter, the most that unemployed, infirm or elderly rural labourers could realistically expect was to be cared for by their children or to gain a place in the local workhouse, the prison-like institutions to which paupers were assigned by parishes. Alternatives, for the able-bodied at least, were migration to one of England's expanding industrial centres or emigration to a far-flung part of the globe. These latter possibilities, however, generally assumed an ability to draw on at least some savings in order to pay for the rail fare or ocean passage and to help in the initial period of settlement in a new town or country where, in any case, one precarious and miserable existence as a labourer would quite possibly be exchanged for yet another.

Although attempts had been made in England in the late 1860s to establish agricultural labourers' trade unions, only in 1871, assisted by a law that gave them legal recognition, did they start to bear permanent fruit.[15] The first of the new generation of agricultural unions was the North Herefordshire and South Shropshire Agricultural Labourers' Improvement Society, but in February 1872 the efforts of this union were overshadowed by an even more vigorous movement among the south Warwickshire labourers, under the leadership of a champion hedge cutter and Methodist preacher, Joseph Arch. There were by now so many scattered local unions in the Midlands and the south of England that some sort of umbrella body was needed to co-ordinate action and protest. In May 1872, Arch's Warwickshire union called a meeting with delegates from twenty-six English counties, leading to the establishment of the first-ever National Agricultural Labourers' Union (NALU). Arch emerged as NALU's first leader and Leamington – a genteel south Warwickshire spa town, until then better known for being "a resort for invalids of mild indisposition, and for wealthy aristocratic sportsmen, and persons of leisure" than as a gathering point of "under-fed, overworked, uneducated [....] voiceless, voteless and hopeless" rural labourers – became its headquarters.[16]

The local and national unions not only sought agricultural labourers as members but village and market town tradesmen and artisans also joined, their circumstances being closely linked to those of local farms. Levels of union backing and militancy varied tremendously in different parts of the country: in no area did all – or even a majority of – labourers join a union which, in many parts of the country, was an insignificant influence. For example, although union activity was strong in east Dorset, it scarcely affected the west, possibly because it had no hiring fairs to cause discontent, and no railways to spread news of campaigns speedily enough. Broadly, the agricultural labourers' movement was most noticeable in the south and east of England, from east Devon and Gloucestershire to Herefordshire, Warwickshire, Lincolnshire and the East Riding, including both pastoral counties and the wheat-growing core of the south and east of the country. Aiming to improve the status and conditions of agricultural labourers, 1872 witnessed a mushrooming of rural unions.[17]

Gloucestershire agricultural labourers, like those in most other counties of southern and central England, shared in the general upswing of rural trade unionism and by April 1872 held meetings in villages across the county. The inspiration seems to have derived initially from activities in neighbouring Herefordshire, but the national union was soon to exert its influence as well. In the early days the Gloucestershire movement appears to have been largely led by farm labourers, but by mid-1872 this began to change when a Gloucester mechanic, William Ebenezer Yeats, started to play a leading role. Yeats' political activities prior to becoming involved with agricultural labourers are unclear except that he was active in political discussions involving "capital and labour", first centring the organizational efforts of his Amalgamated and General Labourers' Union in the Stroud area of Gloucestershire and the northwest of the county. Yeats' union expanded and maintained its independence until May 1873 when it formally amalgamated with Arch's NALU.[18] While most officials at branch level were labourers, most of those who filled district positions – as secretaries, chairmen and treasurers – within the union were tradesmen or representatives of the lower middle classes. Yeats himself was a mechanic; other Gloucestershire officials included a pork butcher, a tea dealer, a shopkeeper and a publican. A reason why so few labourers were union officials appears to be that most lacked even basic education – the great majority were completely illiterate. The nature of the leadership must have inhibited radical demands and action but also encouraged other non-labourers to join the union.[19] Yeats argued that although land and labourers should, by rights, be inseparable, he noted how "scarcely a labourer could be found that had the slightest interest in the land", as landowners had gained possession of their property "by direct confiscation".[20] He was not an especially militant leader, explicitly declaring his opposition to strikes, except when "all other means to obtain an amicable settlement had failed".[21] But he was certainly a flamboyant speaker and his speeches presented an impression of someone sincere in his desire to bring about an improvement in the low wages and poor conditions of employment that farm labourers generally endured. By July, links were being forged between the newly emerged NALU and the Gloucestershire Labourers' Union, and within a month Yeats had become secretary of the latter organization, first as an unpaid official and soon as a fully engaged officer of the national union.[22]

Despite the conditions endured both by union members and by the wider rural population, NALU's aims were always limited, with agrarian reform and land redistribution not on its agenda, campaigning rather under the modest slogan of "a fair day's wage for a fair day's work".[23] Even in the union's heartlands of southern England, militancy was kept firmly under control, with any strike action carefully targeted, something that impressed (and surprised) a reporter with the *The Illustrated London News*:

Hitherto, strange as it may seem, the agricultural strike has been characterised by an entire abstinence from overbearing or violent methods. Those who are conducting it have employed the influence they have obtained with moderation, sagacity, and good feeling. [....] Wherever Arch goes he uniformly cautions his followers that their combination should take the form of 'defence, not defiance;' and that, even if they should possess 'a giant's strength, 'twere tyrannous to use it like a giant.' To act fairly and justly to their employer in striking to raise themselves from their depressed condition is a duty, he tells his followers, from which they must never swerve. To avail themselves of their union at critical periods of the year, such as harvest or when stacks are uncovered for thrashing, would be alike impolitic and dishonest.[24]

An important element in NALU's programme was "to assist deserving and suitable labourers to migrate and emigrate", but while this was not fully formulated until late 1873, independent agricultural labourers' unions, as well as local and district branches of the newly emerged national union, earlier on fully supported, encouraged and directed migration initiatives.[25] The union was involved in the every detail of the migration process, from assisting members to select destinations to their dispatch. Agricultural labourers were often completely ignorant of anywhere beyond the immediate surroundings of the home village and, as was even the case with those who were 'migrated' in 1868 by a sympathetic clergyman from low-wage Devon to northern and southeast England, were often incapable of reaching their destinations without considerable guidance:

> Almost everything had to be done for them, their luggage addressed, their railway tickets taken, and full and plain directions given to the simple travellers. The plan adopted when the labourers were leaving for their new homes, was to give them [...] plain directions written on a piece of paper in a large and legible hand. These were shown to the officials on the several lines of railway who [...] rendered all the assistance in their power by readily helping the labourers out of their travelling difficulties. Many of the peasants of North Devon were so ignorant of the whereabouts of the places to which they were about to be sent, that they often asked whether they were going 'over the water'.[26]

Estimates as to the number of union-assisted English agricultural migrants and their families vary widely, but certainly tens of thousands were sent overseas during the 1870s alone, with estimates for the numbers from 1871 to 1881 ranging wildly from fifty thousand to two hundred thousand.[27]

Internal migration to areas of rural England experiencing labour shortages was at first considered but judged to be impractical as the required skills varied considerably from region to region and, in any case, there was a danger that wages in places where labour was in demand would be driven down by an influx. Far more common an occurrence was for agricultural labourers and their families to turn to cities such as London, Birmingham or Manchester in search of secure work.[28] Overseas migration, in contrast, would remove people from the English rural labour market for what were trusted by the union leadership to be better opportunities abroad, with the intended consequence that the bargaining position, and therefore wages, of those labourers remaining in England would improve.[29] But while Arch considered his mission in Messianic terms of rescuing "these white slaves of England with darkness all about them", people he likened to "the Children of Israel waiting for some one to lead them out of the land of Egypt," he expected the union-sponsored emigrants to be wage-earners abroad, just as they were in England (albeit with better pay and conditions in their new homes) rather than independent producers.[30]

As far as overseas prospects were concerned, Queensland, the United States, Canada and New Zealand each had union advocates at different times, with all receiving large numbers of English agricultural labourers for work on farms, as railway navvies or for similar employment. Articles and advertisements describing and discussing the merits of competing destinations appeared in virtually every edition of *The Labourers' Union Chronicle*, the unofficial newspaper of the national union. Minnesota, for example, was described as "the state where every farmer is his own landlord, and every tiller of the soil is his own master", while Texas, Virginia and Louisiana received similar praise in the columns of the *Chronicle* during the newspaper's first year of circulation.[31] Joseph Arch claimed that he personally had always been greatly troubled at the thought that agricultural labourers were being driven to emigrate by poverty and lack of alternatives at home, later recalling having said in 1872, "I love my country; but I love my countrymen better. The name of a country is nothing to me, if she leaves her honest working men to live upon starvation wages and die in the wards of the workhouse."[32] He came down in favour of Canada ("that they may be Englishmen still"[33]), as the most suitable place for settlement, sometimes speaking alongside Canadian emigration agents at union gatherings in villages and towns across England and even crossing the Atlantic as a guest of the Canadian government in order to make a personal evaluation of employment prospects and conditions on farms in Québec and Ontario.[34] Apart from Arch, Canada had many backers within the union leadership, the country's supposedly healthy climate said to be particularly suitable for labourers from England, although perhaps not an ideal choice for all emigrants: "the worst trade is that of a doctor, for people are seldom sick until they die".[35]

But rather than hindering recruitment, if anything the fact that Brazil appeared so very alien to English experiences could be a positive attraction. The more radical supporters of emigration asked why should English labourers, who were living in constant fear of being forced into a workhouse at home, take an interest in promoting Britain's imperial ambitions abroad? "Is it anything but natural", a union voice was to ask later, "that shipload after shipload of our peasants glide down the Mersey, breathing muttered curses on the land that reared them, and the laws that made their life one ceaseless struggle with poverty? Why should these men love the British flag? What has England and England's boasted Constitution done for those of her sons who toil the hardest of all her children? She has allowed them to be gradually robbed of their commons, and their share in the soil of their own country, by heartless landowners and titled legislators."[36]

Chapter 5
"The Workhouse – or Brazil": The Recruitment of English Emigrants

EMIGRATION AGENTS, SHIPPING and railway companies, British colonial and foreign consular officials all worked closely with union officials in spreading the word of the attractions of their competing destinations. At local and district union gatherings, by personal canvassing at people's homes, and through letters and articles in newspaper columns, agricultural labourers became familiar with Brazil as one of many such potential destinations. Launched in June 1872, *The Labourers' Union Chronicle*, a newspaper affiliated to the National Agricultural Labourers' Union, featured in its second issue an editorial that praised Brazil as an especially ideal destination for English emigrants:

> We are requested to draw attention to an advertisement in this month's issue in reference to emigration to Brazil. This is the only emigration scheme we know of which offers such terms as will enable persons, being agricultural labourers, to emigrate without the means of even paying part of their own passage. The Brazilian Consul-General is empowered by his government to defray the entire cost of emigrants, from their own homes to any part of the Brazilian empire they may select, preference of course being given by Englishmen to the most temperate districts. On arrival, they are at once enabled to settle upon good, productive, well-cleared land, the government advancing them sufficient wages to live upon for the first six months, until a crop is raised, when they begin to make repayment. This they can do by instalments extending over several years, so that the emigrant has an ultimate and not too remote certainty of securing an independent and prosperous position as a freehold cultivator, while in the meantime he is able to gain a satisfactory livelihood.[1]

The idea of securing a farm of one's own and to be paid to work on it must have seemed a quite extraordinary opportunity to a rural labourer who could never even dream of such a possibility at home. It is also hardly surprising that many humble workers could be convinced that the Brazilian scheme was superior to competing options. Canada and New Zealand, for example, were at the time offering merely employment on farms to emigrants who could bring with them little or no capital, offering them no more than a vague prospect of eventual farm-ownership. Most other destinations required emigrants to advance at least a contribution towards the cost of a sea passage, whereas Brazil was providing a loan for the entire fare and paying the difference between the cost of a passage to New York and the longer and more expensive Rio shipping routes. Not even needing to raise any money for the journey to Liverpool, the Brazilian offers must have seemed especially attractive to large families or to those without friends or relatives in other parts of the world who might otherwise have provided assistance.

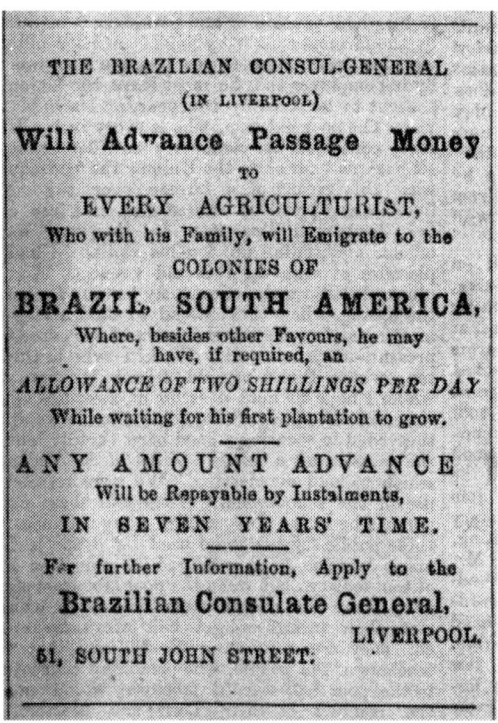

Advertisement in the Labourers' Union Chronicle, 18 January 1873. (Warwickshire County Record Office)

5. "The Workhouse – or Brazil": The Recruitment of English Emigrants

The agents representing Brazil – including Thomas Alsop based in Warwickshire, Edward Haynes in Oxfordshire, and William Ebenezer Yeats in Gloucester, none of whom had any personal experience of the country that they were promoting – were appointed by the Brazilian consulate in Liverpool to work in conjunction with union branches. Travelling widely in their home and neighbouring counties in central and south-western England, the agents set out to persuade rural workers that their future lay in Brazil, a country that most (if not all) of them would have known nothing about beyond the emigration scheme's glowing publicity. Sometimes accompanied by Brazil's Liverpool consul-general, Melchior Carneiro de Mendoça Franco, or by another consular official, Manzilla Meston, the agents attended union meetings to explain the scheme's attractions, aiming to recruit striking – or otherwise disgruntled – labourers, as well as other residents of the localities that they were visiting. On what basis the agents were paid is not entirely clear, an issue that was to be a recurring source of controversy. The agents were always adamant that they were not paid "head money" (payment for each emigrant recruited), insisting instead that all that they received for their services were monthly salaries from the Brazilian consulate, offering no incentive to encourage unsuitable or unwilling people to emigrate as this would not boost their personal income.[2] An understanding, however, of emigration agents' usual terms – under which they were paid £1 or more per person recruited – enabled sceptics of a Brazilian destination for English workers to reasonably doubt this claim:

> a not inconsiderable number of persons who apparently anticipate a lucrative issue to their connection with the various immigration contracts under which the Brazilian Government offers so many pounds sterling for every immigrant whom these companies can induce to embark to Brazil [...] the persons who act as the middlemen or emigrant brokers, and to whom the emigrants themselves are merely a species of merchandize which they purchase in the cheapest market and sell in the dearest, I cannot believe to be deserving of any consideration.[3]

Thomas Alsop was the first emigration agent to recruit agricultural labourers for Brazil and, initially at least, probably the most sincere in his enthusiasm for the country. Alsop, a small farmer and butcher from the south Warwickshire village of Napton-on-the-Hill, became involved with Brazilian emigration in late 1871 when he decided to travel to South America to claim some farmland for himself. Alsop reached an arrangement to travel to Brazil with three young men from his home village where they would work under him on a sharecropping arrangement. It was agreed that Alsop would provide farming implements, seed and horses while Thomas Sheasby, William Rainbow and Arthur Cooke, a nephew of Alsop's, would provide their labour. The profits of the partnership would be divided amongst the four men. As word spread of their plans, other men and families from around Napton became interested in emigrating, with Alsop acting as the agent.[4]

Although Alsop himself does not appear to have been a union official or even a member, he most certainly worked closely with local – and later, the national – unions to seek people who could be persuaded to emigrate to Brazil. Guided by Alsop, during the winter and spring of 1872 Brazilians focused their recruitment drive on Warwickshire, Joseph Arch's home ground, attending meetings in Wellesbourne, Harbury and other villages as guests of the NALU. Typically, villagers would be summoned to union meetings by a man sounding a hand bell and calling out the appointed time and location of the gathering. Although Brazilians were probably the most exotic of guest speakers at such meetings, foreign recruiters were drawn regularly to rural Warwickshire, as elsewhere in England, in competition for would-be emigrants. With discontent running high, they could always attract an attentive audience.[5]

The meeting that the Brazilian consular official Manzilla Meston and Alsop attended on 20 April in Napton – with over nine hundred inhabitants quite a large village in the heart of what was one of Warwickshire's most sparsely populated districts – was typical of such gatherings.[6] Held on the roadside a few metres from the village green and the Crown Inn – which, together with the Anglican church perched on a hilltop overlooking the village, was at the centre of community life – the meeting had been called to discuss the establishment a NALU branch in Napton. After reaching a decision on this, due to the presence of the Brazilian guest, discussion turned to the relative merits of the competing destinations being offered to prospective emigrants. To cries of "hear, hear", Mr J. Durham, the meeting's chairman, asked the assembled crowd what possible prospects existed in England apart from the workhouse and, ultimately, a pauper's funeral? For poor agricultural labourers,

5. "The Workhouse – or Brazil": The Recruitment of English Emigrants

Agricultural labourers meeting in Wellesbourne, Warwickshire, 1872.

Durham asserted, amidst enthusiastic cries of agreement, life in England was nothing more than servility and hardship:

> Drudge when young, drudge when old;
> Drudge through heat and drudge though cold;
> Drudge for clothing; drudge for bread;
> Drudge for the shelter o'er your head.
> Drudge a rich man's purse to fill;
> Drudge the country's mint to swell;
> Drudge the farmer's crops to reap;
> Drudge the banker's grooms to keep.
> Drudge to keep up royal show;
> Drudge to keep off foreign foe;
> Drudge to keep the Church in style;
> Drudge to pay for legal guile.[7]

The future for England's labouring poor, concluded Durham, lay abroad, in New Zealand, Canada or Brazil. Having warmed up the crowd with yet more union slogans and songs, Durham suggested that English agricultural labourers might best successfully achieve "an independence" in Brazil. Invited to explain his country's "emigration system", Meston stated that the Brazilian authorities would advance the fare from Napton to Rio de Janeiro, an amount repayable over a minimum period of two years, according to an individual's personal circumstances. All that was normally necessary on the part of a prospective emigrant was confirmation of being an "agriculturalist" (union membership, involving simply paying 6d in sign-up dues, would do) and a testimony of good conduct from either civil or church authorities. Once in Rio, the new arrivals would be provided temporary lodging in the government's immigrants' hostel before being transferred to one of the state agricultural colonies. There, land would be distributed on what were laid out as highly attractive terms and the settlers would be paid an allowance of two shillings a day – more than most agricultural labourers earned in England – until the first crops were ready for harvest.[8] Meston categorically denied suggestions, made by an anonymous contributor in a local newspaper article that very day, that Brazil was an unhealthy destination, explaining that in the south of the country, where English emigrants would be sent, the climate was "far superior" to that of England and that the land was suitable for the production of sugar, tobacco and coffee, as well as wheat, barley and "all other English crops".[9]

Following the meeting, Meston and Alsop visited the cottages of individuals thought to be interested in emigrating, where they distributed prospectuses and again explained the dual attraction of guaranteed land ownership and of the high wages that awaited English immigrants, tailoring their emphasis on the attractions to the individual on whom they were calling. Many potential emigrants had little interest in land ownership, attracted rather by promises of guaranteed work as farm labourers, high wages and an ample sum to purchase the "beef, mutton, veal, lamb, pork, fowls, ducks, potatoes, and various vegetables" that were said to constitute the Brazilian diet – all great luxuries compared to the monotonous day-to-day diet of most labourers in central and southern England. For example, Thomas Fell, a farm labourer who was earning sixteen shillings per week, was assured that he would earn five shillings a day as well as being provided with housing.[10]

Rather than discussing the realities of life as a backwoods pioneer, the prospectus assured would-be emigrants that all the colonies contained streets, squares, public pleasure-grounds, Catholic and Protestant churches, as well as a school, cemetery, town hall, prison, and other

5. "The Workhouse – or Brazil": The Recruitment of English Emigrants

administrative, social and commercial buildings. Glowing information on Brazil's soil, agricultural produce, physical features, waterways and mines was also included, highlighting five colonies administered and funded by the central government, two of which – Assunguy in Paraná and Cananéia, on the coast of São Paulo – were to be the main destinations of British emigrants. Paraná (and by implication Assunguy) was said to possess diamonds, gold and lead mines, natural pastures for sheep and cattle and "bituminous" earth on which a whole range of temperate and tropical produce could be grown. São Paulo was similarly blessed, with the added advantages of precious stones, copper and silver mines.[11] At best, the information presented was misleading as it appeared that the listed produce was to be found within the individual settlements. At worst, the impressive 'facts' offered were completely false. Paraná, for example, had none of the stated mines, and, as the British consul in Rio de Janeiro was to comment wryly, "whatever hidden treasures the province of Paraná may contain [I cannot] imagine that its possession of 'bituminous earth' can be of any advantage to intending colonists".[12] Furthermore, concerning agricultural production, whatever the natural advantages of both Assunguy and Cananéia may have held out, it was clearly something of a false boast to claim an ability to produce both temperate and tropical crops.

Whereas the farmer Thomas Alsop recruited emigrants with the assistance of union branches in Warwickshire and elsewhere, William Yeats (initially at least) combined his official duties as secretary of the NALU's Gloucester district with a private role as an immigration agent. In early 1872 Yeats first urged labourers to consider "the great benefits of emigration", advocating Queensland, where day labourers for farms were particularly sought.[13] By autumn Yeats had instead agreed to act for the Brazilian government, counselling individuals considering prospects abroad, speaking at small and large public gatherings and distributing leaflets to those able to read:

Emigrate!! Emigrate!!
Important to Agricultural Labourers and their Families

The railway fares to the port of embarkation and passage-money will be advanced immediately to eligible families who are willing to emigrate to the English Colony of Cananea [*sic*], South America. The advanced money to be repaid at the end of seven years. Each member of the family will receive £2 upon their arrival, and 2s per day each, together with the use of tools, seeds, and everything necessary until their first crop is gathered. All willing and useful labourers will find this a golden opportunity, as the numbers will be limited, and every facility will be given to aid them.

Apply at once to

Mr. W.E. Yeats,
District Secretary of the National Agricultural Labourers' Union, Oxford Street, Gloucester.[14]

Other handbills issued by agents provided details of similar offerings, often focussing on the ready availability of excellent and inexpensive food, the emphasis being on meat – beef, mutton, veal, pork, fowl, ducks – especially attractive to labourers used to a simple and monotonous diet.[15]

Despite the natural wealth being boasted of, even at this stage some people were voicing scepticism or even hostility towards the country – "a more mongrel, indolent population is not to be found anywhere" was a typical negative comment, saying more about the writer's prejudices than his understanding of Brazil.[16] But attempting perhaps to avoid controversy, promotional material did not always specifically identify Brazil as the location of a proposed settlement.[17] To a labourer with little or no schooling, for whom it was impossible to distinguish one continent from another, the so-called "English Colony of Cananea" [sic] was likely to have reassuringly suggested some distant part of the British Empire, just another fine place where "poor folks" went, where "wages were higher, and they could live cheaper".[18] It is hardly surprising that, as the emigrants' totality of knowledge relating to Brazil was that provided, directly or indirectly, by the country's promoters, some recruits were later to claim that they were surprised to have found themselves in so foreign a land.

Yeats at first concentrated his recruitment efforts in Gloucester, Cirencester and other Gloucestershire towns, keen to sign-up as many people as possible in the shortest time, giving some weight to the belief that payments were made based on 'head money' rather than on a salary. Numerous surviving testimonies describe the circumstances in which emigrants were recruited by Yeats as he wore down resistance by making wild claims about what awaited immigrants in Brazil. Louisa Bayliss, for example, recalled how her husband, Thomas was approached five times during the course of a single week by Yeats, who even visited his cottage in Gloucester, after he expressed interest in emigrating. Louisa claimed that her husband was told that in Brazil he would earn between £5 and £6 per week (others in Gloucester, such as Albert Lane and Reuben Walker, claimed that they had been told to expect to earn between five to ten shillings a day) and that in Liverpool he would be given money for a new suit of clothes or two suits for each family member on arrival in Brazil. The fact that Bayliss was a blacksmith employed by the Great Western Railway, Walker a timber yard worker and Lane an employee of the Gloucester Wagon Company was not considered a particular obstacle to acceptance in Brazil as farmers: they each paid Yeats six pence to join the union and, now certified agricultural labourers, were coached before they appeared before the Brazilian consulate in Liverpool to claim their passages and passports.[19] Whatever the precise occupational backgrounds might have been of many of Yeats' recruits, they managed to impress George Buckley Mathew, the otherwise sceptical British minister in Rio, who remarked in January 1873 that "the class of persons now coming out are honest, hardworking men, especially from Gloucestershire and Somersetshire".[20]

James Estcourt, another Gloucestershire emigrant, later testified how, when helping to load a cart with his fellow emigrants, he heard Yeats mutter, "I don't care a damn now what becomes of the fellows now that I have their luggage, nor what becomes of them, or whether they go to the bottom".[21] While this might indicate that Yeats was motivated entirely by the head money, it is difficult to imagine that words giving voice to such overt callousness would have been heard (and then ignored) by those setting off for Brazil. Yeats, as Gloucestershire agricultural labourers' union district-secretary, presumably had at least some interest in the well-being of the people he was helping to send to far-off South America. Even if, though late the following year taken to task by the union's national leadership over his handling of unrelated elements of the district's finances, Yeats was simply motivated by personal gain in relation to his Brazilian-recruitment activities, he surely would have kept his feelings to himself.

Despite the failure of the Irish and English immigrants in Príncipe Dom Pedro, news circulating in England of Brazil – "that vast and favoured land"[22] – remained overwhelmingly supportive, with local and union newspapers publishing articles presenting images of a paradise on earth. The *Leamington Chronicle's* picture of Brazilian conditions was typical of the descriptions being offered:

> The climate is genial and regular – without those extremes of heat and cold which characterise the climate of Canada – and the soil is well adapted for the growth of the description of agricultural produce common in England, as well as such tropical products as sugar, coffee, etc., which are beginning to afford so much pleasure and profit to the settler in the colony of Queensland, and which there is a universal demand. The general aspect of the country, is one of exceeding beauty and attractiveness. In some parts it resembles a vast flower garden and almost overpowers the senses by the intensity of its loveliness. In its social and political conditions, the country has made great advances under the sway of its present enlightened and benevolent ruler.[23]

Probably more important as tools to convince prospective emigrants and their friends and relatives were letters circulating that were credited to English immigrants settled in Brazil. Newspapers regularly featured such purportedly first-hand reports, supplied by the Brazilian consulates in Liverpool or London, their English agents and union officials. Reports credited to satisfied Meadows and Christopher recruits who had gone out in 1868 as well as to the familiar names of newly departed emigrants, all of whom were quite overcome by their good fortune, served to endorse the view that Brazil was a supremely desirable destination. "It is very beautiful", wrote Mary Emery, formerly of the Warwickshire village of Pailton, of her new home, Cananéia, "all hills and woods, and there are no trees like there are in England, they are much prettier. The cabbage here grow about twenty feet on the top of the trees."[24]

The prospect of immediate land ownership for agricultural labourers was generally seen as the most attractive feature of the Brazilian schemes and something that came out in many letters. Writing from Assunguy to his brother in England, John Crossley claimed to be overjoyed about his situation as an owner of 75 acres of land, two acres of which were already cleared of forest and planted:

5. "The Workhouse – or Brazil": The Recruitment of English Emigrants

I am as happy as a King, and I don't care a fig for nobody, for I have got a good farm of my own, and nobody to come for the rent like there is in England. We can just work for three months and sow the land, and the next three we can do what we like, then the next three take in our crop and plant another, and then go shooting and hunting. We can shoot plenty of partridges and wild pigs to eat; it is a fine country, anything will grow if you put it in the ground. We have two crops of corn every year.[25]

Despite claiming to have two hundred coffee trees, one thousand grape vines, banana plants, orange and lemon trees "and other different kinds of fruit you never saw" on his 150-acre Cananéia property, emigrant Edward Lyons – a lamp maker from Birmingham – accepted that he would not "make riches" for a year or two, but felt reassured that he would "suffer nothing from poverty or want, as any man here can plant sufficient for their own family". Whatever else the future might hold, Lyons expressed confidence that he would not be "starved out" as, with a bit of work, every kind of food imaginable was readily available, and he was no longer "in dread of the workhouse in my old age".[26]

Writing to his father in Middleton Cheney, Northamptonshire, Alfred Merry also told of the happiness that he had found in Assunguy (which he called "the healthiest place in Brazil"[27]), feelings that extended beyond merely material well-being, satisfactory though that was:

The Government has behaved well to me since I have been here. They have given to me an axe, a billhook, a knife (two feet long) a boiling kettle, a large saucepan, ladle, a spoon, plate, cup, and they have been keeping us from the 29th of June to the 3rd of October, and have not asked me for no money, but gave me one £11. We can go where we please – do what we please. *This is a free country*. The natives of this place are so kind. Every time as you pass their house you must go in and eat and drink. They are pleased to see us English people; but they do not like the French.[28]

Some letter-writers were more realistic in explaining their good fortune as backwoods pioneers. In November 1872 Edward Young, from London, wrote proudly to friends that, just a year after arriving in Cananéia, he was the owner of 150 acres of land and that his father held a further three hundred acres:

113

We have 14 acres cut down, and two of it cleared of all except tree stumps, and my house is ready for the roof. The crops here yield well, but all English things will not grow well here – potatoes for one thing – but as there are six or seven [other varieties of tubers] here that is of no importance. Onions do not seed but are parted, the same as garlic and planted, and they grow well. Radishes grow very well here, the English especially; we have some now three or four inches in diameter, And as solid as any turnip. Scarlet-runners do not grow well, and what wheat we have planted has not come up. The Indian corn and rice grow splendidly here, and you will make flour almost equal to wheaten flour when they are mixed together, and if you can bring any curry, you could have a good meal of rice and curry at any time.[29]

But even during the peak recruitment months of 1872, critical references to Brazilian conditions entered into some of the letters of even apparently satisfied settlers, including occasional references to some of their fellow emigrants being entirely discontented. If anything, however, this served to increase the apparent veracity of other parts of the letters that extolled the settlers' new surroundings. The Birmingham-emigrant Edward Lyons, though claiming to be generally satisfied with conditions in Brazil and optimistic about his future, mentioned that the recently-arrived Oxfordshire families in Cananéia were "a useless lot", unwilling to work and poorly thought of, attracted to Brazil in the expectation of eating fresh beef, mutton, "and every other nice thing for nothing".[30] J.W. Apperley admitted that he found Assunguy to be in some respects better than expected, but in others conditions were worse, and he explained that "it's no good for anyone to think of coming out here unless he means to work and to rough it for the first two years".[31]

But in spite of hints of discord, Brazil was said to offer hopes for a secure future, free from hunger or fear of it, to those willing to work hard. "I do not expect to make riches here for a year or so," wrote Lyons in a letter to his father-in-law with the explicit purpose of convincing him to join his family in Cananéia,

> but one thing I am confident of, I will suffer nothing from poverty or want, as any man here can plant sufficient for their own family as far as good substantial food is concerned, I mean pork, corn flour, potatoes, eggs, fowl, milk, coffee and sugar; also if I want a glass of

5. *"The Workhouse – or Brazil"*: *The Recruitment of English Emigrants*

country rum you can purchase it at 2s [10 pence] per gallon, so that I am in no way afraid of being starved out, nor neither am I in dread of the workhouse in my old days. I fear no landlord or agent turning me out for the rent, as house and land is my own.[32]

In urging his father-in-law to join him (and telling him to expect a visit from Brazil's Warwickshire agent, Thomas Alsop, who would arrange passage to South America), Lyons was clear that one must be prepared to rough it for five or six months, but that a veritable Garden of Eden was awaiting English emigrants. Of his 150 acres, Lyons claimed to have forty acres clear of forest on which was planted corn, rice, sugar, arrowroot, several species of potatoes, tobacco, garden vegetables, coffee, grape vines, bananas, orange and lemon trees and "other different kinds of fruit you never saw".[33]

Last sight of home: setting off for a new life abroad. (Tate Gallery)

Whether these letters were entirely freely written (or dictated – many of them were from people who were illiterate, marking their names with a mere 'X') it is impossible to judge with any certainty. Certainly the individuals credited with writing the letters actually existed and even later, when the Brazilian emigration schemes were coming under severe attack, there were few direct suggestions that the letters might have been penned on their behalf by people with an interest in promoting emigration. But according to Thomas Alsop himself, to save postage, letters from Cananéia and Assunguy were sent to the Brazilian consulate in Liverpool and on to him at Napton. Alsop would then deliver the letters personally and make copies to circulate to newspapers. One can only speculate whether letters containing unfavourable accounts were withheld by the consulate or by Alsop, or whether Alsop re-wrote letters as he copied them.[34] Commenting on a series of entirely favourable extracts from letters that were reproduced in an issue of *The Labourers' Union Chronicle*, a Foreign Office official simply noted "[i]t is curious if the prospects of the Colony were so brilliant that [the immigrants] did not remain there".[35] If, as is certain, there were exaggerations (or even complete untruths), these can be explained by a desire on the part of the letter's sender either to reassure folk back home that they were safe and successful or in hope that they could tempt loved-ones to join them. Edward Lyons, the Birmingham lamp maker, wrote that he "would wish to see a thousand countrymen here", to offset the Catholic dominance in the colony and urged his father-in-law to consider emigrating, advising him that Alsop would soon be calling.[36] In terms of content and style, the letters conform in authenticity to the general utilitarian pattern of nineteenth-century emigrants' letters.[37] The basic details of what to expect in Brazil are cited again and again, with little variation: excellent land available on easy terms, cheap and plentiful food (especially meat), hunting free to all, a beautiful and friendly land of natural abundance where security is assured through old age, and an idyllic climate where, in the words of the Oxfordshire-based agent, Edward Haynes, "the sun shines on both sides of the hedges".[38]

Chapter 6
From England to the Brazilian Colonies

WITHIN WEEKS – OR even just days – of being signed up as recruits by agents representing Brazil, individual emigrants, as well as families, set off from their home villages or towns in England. In most instances, the first stop was Liverpool, where, at the Brazilian consulate, the emigrants would formally declare themselves to have experience working on farms (which many had, but others certainly had not), membership of an agricultural labourers' union being considered as reliable evidence.

Even the best organized and financed pioneering venture would involve hardships but, as has been (or will be) seen in relation to the Brazilian state colonies that received the vast majority of British immigrants, the conditions experienced in Brazil were especially harsh. But hundreds of English would-be pioneer settlers did not make it beyond Rio de Janeiro's immigrants' reception centre, the Casa de Saúde, while many others, if they did leave the capital, experienced weeks or months in hastily arranged transit centres or temporary lodgings, waiting for land to become available for allocation. For these immigrants, to be so tortuously close to what should have been their new homes must have heightened their feelings of frustration, and the overcrowded and otherwise appalling conditions that they endured made them particularly vulnerable to disease and even death. With hopes dashed after barely taking their first steps in Brazil, often with the most tragic of consequences, it was inevitable that negative reports regarding conditions (and the apparent deception of the immigrants) would soon circulate in their home villages and towns.

Thomas Alsop, along with the first party of 114 men, women and children that he and Edward Haynes had recruited in Warwickshire and Oxfordshire, sailed from Liverpool on 29 May 1872 on the Pacific

Steam Navigation Company's ship, the *Lusitania*. Also on board were parties of emigrants from Wales and Derbyshire, a joint Swedish–British team intent on surveying a proposed transcontinental road and a number of Brazilians returning home. After a brief call at Bordeaux, the *Lusitania* docked at Lisbon to take on additional cargo and passengers. There, Alsop and some of his group went ashore and spent a day wandering around the city, for all of them their first experience of a foreign country. With most of this group coming from farming backgrounds, it was the market and its range of agricultural produce that they found especially astonishing: apart from seeing fruit, such as oranges, that were completely new to them, they were amazed that produce they were familiar with, such as cherries and peas, was available so early in the year. Soon the voyage resumed, but fortunately the Atlantic crossing appears to have been uneventful: many of the women and children suffered from sea sickness, a baby was born, but for the most part the passengers simply relaxed, passing their time below decks eating, drinking and dancing, or watching out for dolphins, flying fish and whales.[1]

Early on Sunday, 19 June – just three hours short of twenty days since leaving Liverpool – the *Lusitania* entered Guanabara Bay, dropping anchor offshore from the city of Rio de Janeiro. The *Lusitania* was immediately surrounded by a swarm of rowboats, and soon a steam tug appeared alongside to take Alsop and his party ashore. There, two English-speaking immigration officials (a German and an American) met the group and, after enquiring as to how the passengers had fared during the voyage, escorted them to the Casa de Saúde. There was much commotion at the hostel as the English and other *Lusitania* passengers were settling into their temporary home just as a large group of Germans were being transferred to their chosen or assigned colonies. Having settled into their accommodation, the Casa de Saúde's Alsatian administrator formally welcomed the party and at once arranged a hearty dinner to be served consisting of soup, beef, bread, rice and beans.[2]

During the group's first evening in Brazil, the Ministry of Agriculture's agent of colonization visited the hostel, accompanied by several British residents of the city who, as before, were curious to meet the newcomers.[3] After meeting with the agent to discuss the relative merits of the available state colonies, Alsop and his group decided not to proceed to Santa Catarina, as had been their original intention,[4] instead opting for Cananéia, on the coast of the province of São Paulo, some three hundred kilometres southwest of Santos. Alsop later explained that while in Rio he had met two British immigrants who had just returned from Santa Catarina and had warned him about the supposedly disruptive behaviour of German immigrants there.[5] It is also possible that news had reached the group of the recent failure of the English and Irish settlers at Colônia Príncipe Dom Pedro; William Scully's altogether negative views of Santa Catarina may also have been influential. Whatever the reasons for the change, Colônia Cananéia, located as it was near the coast, was certainly made to sound a more attractive destination. The colony, the group was reassured, was directed by an anglophile – Elpídio de Mello, who had been educated partly in London and was married to an English woman – who planned to populate the colony exclusively with British immigrants. Several single young men decided not to proceed to Cananéia, however. Some of these found work with an engineering expedition while others, as had also been the case with several families from the Banbury area, chose to accompany a largely Welsh group (fellow passengers from the *Lusitania*) north to Piúma in the province of Espírito Santo to settle as sharecroppers on the Fazenda Montebello, a large British-owned coffee plantation being developed as a private agricultural settlement.[6]

Despite the warmth of the reception, Alsop – who had never before left England – was at once struck by what he regarded as the poor organization of the "Brazilian emigration business", a disorder that he considered was compounded by frequent changes of government ministers, directors and immigration officials who, though "very kind and obliging [....] do not conduct their business in quite so English a style as I had been accustomed to".[7]

While in Rio, the immigrants spent their days wandering the narrow streets of the city, surprised that "only the nobs" wore shoes, examining the shops and markets, comparing English and Brazilian prices and discovering in the market stalls what was, for them, new and exotic produce. For some, the greatest shock was that Portuguese was spoken in Rio de Janeiro and even surprise at being in Brazil at all. "I thought I was going to America where they could talk English," wrote one young man to his mother, "but no, you have got to learn their talk."[8] Eager to downplay the feeling of frustration from the prolonged stay in the Brazilian capital, Alsop later claimed that his party was only delayed by ten days and that this time was enjoyably spent, "visiting the principal places of interest in and near the city, with which we all were highly delighted."[9] Nevertheless, Alsop accepted that some discomfort was experienced by members of the group, including an outbreak of smallpox in the hostel after seven days in the city (which he insisted was introduced by immigrants themselves and not contracted from Brazilians), resulting in the death of a child to the "general discouragement" of the party.[10]

Alsop's main complaint – one that he felt could easily have been avoided – was the lack of a connecting vessel to take the party to their new homes. With no regular shipping service linking the capital with Cananéia, there was an enforced three-week stay in Rio until the monthly steamer to Montevideo (which was prepared to call at ports in the south of Brazil) was due to depart.[11] Recognizing that his party was getting increasingly restless, Alsop told the Brazilian officials that, in future, departures from England should be arranged to coincide with the schedule of the coastal steamer and that the stopover in Rio should last no more than a few days.[12] Constantine Phipps, a British diplomat based in the city, agreed, noting that, "this enforced idleness, after the long sea-journey, was apparently the first false step, as being calculated to wean the emigrants from any active and industrious habits they may have had".[13] But this was a problem that the Brazilian authorities were well aware of, themselves complaining that telegrams detailing the impending arrival of parties of immigrants did not always arrive.[14] Finally, on 8 July Alsop was told that the boat for Montevideo was preparing to depart. Two days later the immigrants set off for Cananéia.[15]

In contrast with the *Lusitania*, conditions on board the coastal steamer were cramped and squalid in the extreme – "little better than the black hole of Calcutta", recalled one of its passengers.[16] Instead of calling at the town of Cananéia, located on an island, a special arrangement was made for the steamer to call at the colony's port on the mainland, where Alsop and his group arrived late on the Friday night. With a certain amount of timidity, passengers transferred into canoes to be brought ashore and were provided with accommodation in what Alsop rather grandly referred to as "Government House" – in reality nothing more than a large, hastily erected storage shed on a small bluff surrounded by thick forest. Nevertheless, universal relief was felt amongst the immigrants as they disembarked and looked forward to taking up their new farms and beginning new lives.[17]

Early the following morning, Alsop, the colony's director and his assistant set off inland towards the colony. Over the next few days, Alsop met (to gather information on local life) some of the sixteen British families who had arrived in Cananéia four years earlier. In a letter to *The Labourers' Union Chronicle*, Alsop claimed that all of these settlers were, with but one exception, "perfectly satisfied" with life in the colony, and that the only possible improvement they could think of would be for friends from England to join them.[18] Alsop claimed that he was thoroughly impressed with all that he saw of Cananéia, describing the immigrants' happy circumstances and prospects for new arrivals:

> A tract of land is reserved for village lots, building purposes, and a common. The land is very rich here, and as the traveller advances into the interior of the county [....] he will find several English farmers with large tracts of land, on which grew last year some of the finest Indian corn I ever saw, – rice, beans, peas, potatoes, sugar cane, pine apples, oranges, lemons, etc., and nearly all kinds of vegetables, while some of the land is planted with the grass of the country, upon which cattle thrive tolerably well. Pigs are very cheap: you can purchase a small pig 12 to 14 week's old for a milreis – 2s. 1d. A cow £4 to £5; a good horse or mule, £7 to £8. Poultry is very plentiful, consequently cheap. The director of the colony informed the emigrants that they were not obliged to take their lots in what is called the Colony Proper, but if they preferred to take land outside the boundary, they were at liberty to do so. Each male of the age of 18 years and upward could take from 75 to 300 acres, – that is, from a quarter to a whole section. Most of them said 'we will take the larger quantity, for if we never want it our children may'.[19]

Although Alsop had originally intended to take land in Cananéia, just a few days after arriving in the colony he boarded a southbound steamer. The reason he gave for leaving his party behind was that he wanted to compare Cananéia's qualities with those of the state colony of Assunguy, situated in the neighbouring province of Paraná. Alsop arrived at the port of Paranaguá on 19 July from where he made his way to Curitiba, Paraná's small capital, and onwards to Assunguy, located one hundred kilometres further north. There he met with the British settlers whom he found "generally prosperous" and with one exception (just as before in Cananéia) "perfectly satisfied". Alsop reported Assunguy's land to be preferable to that of Cananéia, but he acknowledged that poor communications and the colony's lack of a market raised problems. Nevertheless, he argued that good opportunities existed, pointing to the example of Samuel Hawgood, a *colono* from London, who had been in Assunguy for several years and supposedly had made a £200 profit from his land the previous year alone. The supposed prosperity of the British *colonos* was all the more remarkable given that most were, like Hawgood, from English towns and cities, suggesting to Alsop the immense possibilities open to settlers who already had agricultural experience.[20]

Saying that he was entirely satisfied with the possibilities that Brazil had to offer, Alsop soon returned to England to recruit more people. According to an article in *The Labourers' Union Chronicle*, Alsop claimed that while in Rio he had had several audiences with the emperor who was especially keen on attracting immigrants from Warwickshire "to teach his subjects 'good farming'".[21] Every advantage would be provided and support would be given to new arrivals, each of whom would become an "independent landlord farmer". Alsop gave the assurance that the cost of Brazilian land was purely nominal, while some of the soil contained "gold, lead, and other minerals" (a reference to the misleading leaflets being distributed in England); he claimed that he had seen during his travels "tons of almost pure lead dug up and abandoned". In truth, seeing the recruitment of emigrants for Brazil as a fine business opportunity, Alsop embarked in November 1872 on a tour of England "to send off the ejected sons of the soil".[22] Perhaps because of the negative accounts now beginning to drift back to Warwickshire and Oxfordshire, Alsop went further afield in quest of recruits, concentrating his attention on the West Country. One of his many stops was the village of Piddletown, near Dorchester in Dorset, where, aloft a wagon, he spoke of the attractions offered by Brazil. Now able to declare that he was speaking on the basis of his own first-hand experiences, Alsop repeated the claim that the agricultural colonies all had churches, chapels and schools and he assured his audience that the three days a week of paid work that the Brazilian government guaranteed immigrants was sufficient for a family to live on comfortably. In a matter of weeks another party of emigrants was assembled and ready to set off.[23]

But as this latest party of emigrants were leaving England, more news that all might not be well with those who had travelled out in May with Alsop was circulating. On 28 July 1872, shortly after arriving in Cananéia, William Brown, of Shotteswell near Banbury in Oxfordshire, feeling totally deceived by the promises made him of what to expect, sent a letter to his parents (passed on to their local newspaper, *The Banbury Guardian*, for publication) imploring them to tell his brother not to join him:

> The houses are not as good as they are in England. We have nothing but black beans and boiled rice, there is no bread here. We have no bed here but the cold ground to lie on. I have not undressed but three times since I left Old England.[24]

Just a few weeks later, 23 August 1872, Mary Stanton, of Great Bourton, near Banbury, who had also joined Alsop after being recruited by Haynes, wrote a desperate letter on in which she described her experiences. After receiving and publishing so many glowing accounts of life in Brazil, a frankly puzzled *Labourers' Union Chronicle* reproduced on 21 December the following letter:

> My dear Mother and Father,
> I write these few lines to you, hoping they will find you quite well, as they leave me not v well nor my child, for he has had the small-pox very bad. He has been blind five days with them; they are turned, but the doctor thinks he will not get well. They have used us very badly. We have not tasted bread since we have been here, we live on rice. They have not put us on any land yet. We live together like pigs – worse than pigsties; and we shall all soon be pined to death here. We would not have come here for all the world if we had known. [....] There are 114 out here with us and we have petitioned the Emperor for us all to go back. If we can, William could work his way. [....] There is no church nor chapel here, nor any parson, there have five died in three weeks. Joseph Goode's son died next to us. Jane has had the small-pox, and she is v ill now. We are in v gt trouble here. You would no know Will now, he's got v thin; he troubles so over me and the child. I have been v ill with the diarrhea, but I am a little better. [....] Willie was such a nice boy before he was ill; but it will break our hearts now to lose him, bless him. Don't fret

about us, for if we don't meet you again on earth we hope to meet you again in heaven. [....] We had to walk twelve miles, up to our knees in mud, without any shoes or stockings on, Will and me, and we had to lay out all night, and it poured down with rain, and it thundered and lightened all night very bad. We laid out in the open air all night, and had to walk on the next morning with our wet clothes on six miles. It was Sunday morning. God guarded us from the wild beasts that roared around us. Dear mother, ask John to show Haynes this letter, and ask him if he is not ashamed of himself to get people out here to be used like this. The people would have killed Mr Alsop if he had not run from here, for bringing them out here like this. Will send his kind love to you, but he troubles so I am afraid he will break his heart over me; he is so v kind to me here. He is wasted to nothing. Dear mother, you asked me to send you the truth, and I have sent it to you, and you can shew anyone this letter you like. Write back again as soon as you can.

Mary Stanton

Dear Mother, I must tell you my baby is dead. He died on Saturday morning at eight o'clock.[25]

But Alsop claimed to have received letters from emigrants "who advise their friends to come to the land of Canaan" and he dismissed his critics as being "dogmatic", insisting that Brazil was "the *best place* to go to" and asking why the Brazilian government would be investing considerable sums of money to assist English farm labourers to settle in agricultural colonies if there was no prospect of them doing well. "It is absurd to say one or two [emigrants] have failed, or that fevers are known at Rio, or that there are snakes. Who would recommend a foreigner to settle in England if he was to be guided by our fogs and our pauperism?" He insisted that the tide of emigration to Brazil would not be stemmed by the "grumbling of three Oxfordshire men", naming the ring-leader of the malcontents as "the returned butler Stanton".[26]

Other reports contradicting Alsop's wholly positive account of his visit to Brazil were published, placing him on the defensive. Refuting accusations, such as that made by Mary Stanton, that he had abandoned his group and fled Cananéia, Alsop explained his decision not to remain in the colony as being simply due to the fact that the men who agreed to work for him had changed their minds when they found that they could set up on their own.²⁷ Such reversals of his opinions regarding Cananéia puzzled the Casa de Saúde's administrator, F.A. Fritsch, who told Phipps, the British consul, that he recalled having been told by Alsop on his return from the colony that it was quite unsuitable for British immigrants. "How was I surprised," pondered Fritsch, "when I saw vessels come from Liverpool loaded with emigrants for Cananea [sic] engaged by Mr Asop."²⁸ Alsop nevertheless continued to insist that conditions in Cananéia were ideally healthy for British immigrants, claiming that before the July arrival of his party there had been only one death in the colony over the course of four years, an unfortunate event that he attributed to an "over indulgence in native rum".²⁹

Apparently oblivious to the mounting controversy surrounding the plight of the emigrants taken in May to Brazil, on 29 November 1872 a further party made up of three hundred people recruited by Alsop and Yeats (but not accompanied by either agent), set off for South America on the *Santa Rosa,* a ship belonging to the Pacific Steamship Company. In contrast with some other voyages, reports refer to the Atlantic crossing as a terrible one, with shocking conditions on board. The ship was not designed to carry emigrants, sleeping accommodation was improvised, with at least six people obliged to eat, wash, sleep and dress within a space measuring just two by a little over one metre. The space between bunks was extremely cramped with the lowest level being just a few centimetres from the floor and often flooded by water and waste swept downwards from the deck. Passengers complained that during the entire forty-day voyage they were unable to change their clothes due to a lack of privacy, as "married and unmarried, men and women, were mixed up together contrary to all ideas of common decency and thereby causing much irregularity and ill-feeling among the people". In the ship's slaughterhouse – livestock were carried on most lengthy sea voyages – animals were butchered within an arm's length of the berths: sheep and other animals were kept almost in contact with people when lying in their bunks. While it was agreed that food was plentiful, people complained that it was badly cooked and that it most certainly did not match the extravagant "bill of fare" printed on the passengers' tickets.³⁰

The immigrants arrived in Rio de Janeiro on 23 December 1872 and spent only about a week at the Casa de Saúde before 245 of the immigrants were transferred to Cananéia. Again, the colony was entirely unprepared for the arrival of such a large party of immigrants and accounts of their reception closely match the scene that had greeted Alsop and his contingent earlier that year. Disembarking at the port early on 3 January, the immigrants were crammed into the four-by-eight metre storage shed where they would take turns sleeping on the dirt floor for the next twenty-four days. Not until the following day was food organized and over the next few weeks their diet was composed of a limited quantity of *carne seca*, (dried beef), black beans and rice, with many of the men finding themselves forced to sell their jackets and coats "to give food to their wives". After a few days, the single men went to the colony to work but as there were neither tools nor food for them there they returned to the port and were immediately ordered back to Rio since it had been decided that only married couples would be given land. Shortly after, twelve married men walked the twenty kilometres to the colony carrying a petition demanding that the entire group be returned to England, a request to which the director – struggling to provide basic sustenance to the new arrivals – was only too pleased to agree.[31]

With each passing day, more and more of the immigrants were falling ill from dysentery and similar afflictions. With the next scheduled steamer to Rio only calling at the town of Cananéia, all the remaining members of the *Santa Rosa* contingent had to trek from the colony's port to the other side of the island. Thomas Bayliss, a Gloucestershire man, had to be carried the entire distance from the colony, where he had taken ill, to the town, while sick children and women had to walk. Bayliss was eventually taken on board the steamer and despite appeals from his fellow passengers, he was placed in the lower deck where water flowed over him. On arrival in Rio, Bayliss was taken to the hospital where he died soon after, one of thirty of the immigrants (three men, a woman and 26 children) who failed to survive.[32]

To protect the returnees from Rio's often intense early February heat when outbreaks of yellow fever were frequently at their most virulent, as well as to keep the party away from newly arrived immigrants from England who would certainly be concerned for their own welfare if they were to hear stories of the conditions experienced in Cananéia, the survivors were sent by rail to Mendes, a hill village in the interior of the province. What to do with returning immigrants – individuals, small or large groups – was becoming a considerable problem both for the cash-strapped (and embarrassed) colonization agency of the Ministry of

6. From England to the Brazilian Colonies

Agriculture, and also for the British business and diplomatic community in Rio who were finding it increasingly difficult to raise funds to support their distressed compatriots. Officials of the colonization agency suggested that the families being held in Mendes should be transferred to Santa Catarina, an offer that British diplomats recommended the immigrants to accept. Already disillusioned with Brazil, the immigrants would not believe the assurances that the mainly German colony of Dona Francisca was capable of absorbing them, any better than Cananéia had been able to. Instead they demanded to be sent to the United States, in all likelihood the original destination of preference for many before they had been tempted by the South American package.[33]

No sooner had William Yeats' *Santa Rosa* contingent left Liverpool than the Gloucestershire agent was busying himself assembling another party of emigrants, again concentrating his activities on the market towns of his home county. Every recruit had, of course, a unique story to tell of departure from England and experiences of Brazil, but the case of George Bond features many common elements. Born in 1828, Bond, a young shoesmith from Cirencester, married Emma Gegg of Rendcomb in 1851. The couple settled in Bond's hometown and there he gradually progressed in his trade and became a master blacksmith. A year after marrying, Emma gave birth to Clara, the couple's first child, followed by five more children over the coming years. The last of the children was born in 1870, their mother dying shortly after. Perhaps because George was unable to cope with Emma's death, the six Bond children – along with Bond's 85-year old mother – were taken to the Cirencester Union Workhouse. Apart from the fact that Bond left his Park Street home and workplace, his whereabouts immediately after he abandoned his family are unknown.[34]

George Bond reappeared in Cirencester to claim his children in late 1872, around the time of his mother's death. By then there was an addition to the family, a baby girl named Lucy, born a workhouse 'inmate' to the unmarried Clara in September. At this time William Yeats was recruiting would-be emigrants for Brazil and somewhere in the county he caught Bond's attention, apparently persuading him, according to Clara, that "we should do a [great] deal better in [Brazil] than in England and should have a [great] deal of money". For a large, destitute family, Brazil provided attractions beyond the chance to start completely afresh: on a most basic, practical, level, there was no requirement that emigrants should contribute up-front towards the cost of their passage, something that Bond would not have been able to do for his large family. Yeats enrolled Bond as a union member and now, as a bonafide agricultural labourer, he was eligible to be taken on as a Brazilian immigrant. Like many of Yeats' recruits, Bond was clearly not an agricultural worker nor even, as was the case with many others, from a farming background. But, as a blacksmith, at least he possessed a skill that would be highly valued by any farming community.[35]

The Bond family left England at Christmas 1872, travelling to Brazil via Liverpool aboard the *Patagonia*, arriving six weeks later in Rio de Janeiro. With no onward transport organized for them and with other newly arrived immigrants occupying the Casa de Saúde, the Bonds, along with most of the other immigrants who had travelled on the *Patagonia*, were sent to Barra do Pirahy, a railway junction located in the province's hilly coffee-growing region, away from the overcrowded and unsanitary conditions of the city of Rio. During the time that the immigrants spent at Barra do Pirahy they were accommodated in a large warehouse belonging to an important local landowner, Commendador Faro – conditions that were basic but ones that the visiting British consul considered to be, as should be expected, superior to those of the several hundred Portuguese immigrants who were being held nearby.[36] Even if judged adequate, arrangements in Barra de Pirahy for the immigrants, now shunted out of sight of the British and Brazilian officials in Rio, found many of the would-be *colonos* unable to tolerate the uncertainty of what was to become of them next.[37]

Aware that there were now several hundred inactive English labourers so close to the city of Rio de Janeiro, Thomas Dutton – one of the partners of Yates & Company, a firm that was developing a coffee estate in the province of Espírito Santos – visited Barra do Pirahy to try and convince some of their number to consider the Colônia Montebello. The basic terms were that the immigrants would be provided with land to be planted with coffee trees, and the crop would be marketed by Yates & Company on whose land the *colonos* would also be obliged to work one day a week. Feeding on the anxieties of the immigrants, Dutton claimed that the Brazilian government was unable to receive them in the state colonies and claimed that there were already six hundred British immigrants idle in Curitiba and elsewhere in Brazil. While interested,

6. *From England to the Brazilian Colonies*

Fazenda Montebello, Espírito Santo. (Biblioteca Nacional, Rio de Janeiro)

some expressed concern that they would be breaking the terms of the contract upon which they had been brought out if they accepted Dutton's offer, while others had been given warning of conditions in the colony. Dutton painted a glowing image of what he claimed lay ahead. "The fertility of the soil and healthy position of Montebello cannot be surpassed in any other part of Brazil" he boasted, in what by now must have been familiar language to the immigrants. Most tempting of all, Dutton was keen to stress that anyone accepting his offer could set off at just five or six days notice and that they would reach their destination within only another 48 hours. If, on the other hand they were so foolish as to remain at Barra do Pirahy, they would have to wait for months until moved on by the government, in the meanwhile remaining "most probably in total idleness, wearing out their clothes and boots and acquiring habits of laziness and immorality".[38]

It is unclear whether or not any of the immigrants at Barra do Pirahy were taken to Montebello, as a few families in Alsop's party had been the previous year, but after a month under the care of Commendador Faro, the Gloucestershire and other English immigrants were returned to Rio and immediately boarded coastal a steamer to carry them on the next leg of the journey towards their expected new homes in Paraná.[39]

Despite having been isolated outside Rio for an entire month, in part to avoid contracting yellow fever, on arriving at the port of Paranaguá the English immigrants were detained at the Ilha das Cobras. This was a quarantine station consisting of rough calico tents with a coarse mat laid on the floor which provided only the most limited shelter from the intense summer rains; water entered in torrents and the ground turned into a quagmire. After ten days – a lengthy period of time, given the conditions, but a short stay compared with the five weeks endured by the *Santa Rosa* party who overlapped with them – the immigrants were taken by boat to Antonina and then onwards to Curitiba by way of the steep road clinging to escarpment on the edge of the Serra do Mar. As Assunguy was still unprepared for the party, the group was placed in the Bariguy Lodge, the provincial government's main immigrants' reception centre, located five kilometres north of Curitiba. By all accounts conditions at the lodge – in fact a collection of hastily erected wooden barracks approximately ten to twenty metres long and five metres wide, with eight families or more squeezed into some – were difficult in the extreme. Deaths were commonplace, infections spread rapidly due to the overcrowded conditions in which as many as 370 English immigrants were concentrated.[40]

As the days waiting at Bariguy turned into weeks and months, the group became increasingly disgruntled with their situation, confused as to why they were being held there and most likely alarmed by the stories that would have been seeping out of Assunguy concerning conditions in the colony. As Assunguy allotments became available for allocation, some of the families were dispatched to the colony. The men travelled the hundred kilometres on foot, the women and children being carried on mule back. A payment of four milreis was given to each settler for food along the way. For most of the immigrants being held at Bariguy, the wait was a long and demoralizing one.[41] Some of the men, at least, were fortunate in being given occasional days or periods of paid employment ("the best work I had in the country", according to one[42]), working on the road intended to link Curitiba with Assunguy, a seemingly endless construction project. Because of his skills as a blacksmith, George Bond was sent to sharpen the tools being used by the construction gangs, leaving the children at the Bariguy Lodge under the care of Clara. But just a few days after leaving the lodge, George Bond fell ill and two weeks later, on 16 May 1873, he died, away from his family.[43]

News of the plight of the now orphaned Bond children eventually made its way back home, with a series of articles appearing in *The Wilts and Gloucestershire Standard*, Cirencester's local newspaper. A letter in the *Standard* from Jane Lander (a woman originally from near Cirencester but then living with her family in Antonina) featured the first passing reference to the Bond orphans, mentioning that the "head man" had "sold some of the children", a sale that was only stopped due to the intervention of other English immigrants.[44] A fellow immigrant from Gloucestershire, John Frederick Davis, who arrived home in Cheltenham on 14 September 1873, expanded significantly – and alarmingly – on the story:

> There was the case of a man named Bond, from Cirencester way. He had no wife and eight children. Well, he heard of some work to make a new road, and went there. He died, and his mates tried to bring him to Curityba, where the rest were buried. They put him in an open horse trough, but the road was that bad and rocky and they kept falling with him, and the last time he fell they could not put him in again, and we[r]e obliged to take him a little aside and bury him with their hands. They put a little cross up, but afterwards another of his mates coming down to Curityba, went to see who was buried there, and found the body exposed and half eaten by wild animals. So the poor children were left orphans, and the men came to take them, as they said, to an orphan asylum. But there was no such thing, sir, and I believe they were sold, for one of the children said they had been sold to those they were with at 5 milreis apiece. Well, sir, the wife of a man named Hutchings had died, and a fearful death too sir, – and at last one of the eldest girls of the Bond family, sir, starving as she was, was induced to go with Hutchings up to the colony as his wife. But there, sir, if you had seen what she and the others suffered, poor things, you would pity her. It was perhaps the only thing that could have saved her from starvation, although you may well say, the saving was little better than that. There are many more sad things I could tell you, sir; but I hope that I have told you enough to make you warn others against being deceived into going out there.[45]

The news that Gloucestershire children – Charles (then aged 13), Annie (11), Emma (10), Ernest (7) and Alfred (3) – might have been sold into white slavery would certainly have shocked, though perhaps not entirely surprised, readers of the *Standard*. A steady flow of letters highlighting the dreadful conditions being experienced was by now reaching England and tales were also being circulated in person by newly returned emigrants. Furthermore, one of the few facts about Brazil that was widely known to the British public was that slavery was legal there, something that made the story seem entirely plausible. George Harmer, the *Standard*'s editor, took it upon himself to contact the Foreign Office to press for an investigation – "it is easy to see how a very flimsy pretence of taking charge of the children might, in the absence of parents and friends to look after them, become in Brazil virtual slavery." Harmer was, however, prepared to retain something of an open mind as to what precisely had become of the children. That said, he felt that "the position of these orphans is pitiable enough, especially when we read of the privations undergone even by the adult Emigrants who are able to take care of themselves". He launched a fund to rescue, as he saw it, the children and to provide them with support on their return home to Cirencester.[46]

But within weeks, the story altered slightly, with a hint that the children may even have benefited from their father's death. "George Bond, a confirmed drunkard," the Foreign Office informed Harmer, "died near Curitiba of *delirium tremens*, and his eldest daughter who had an illegitimate child, went off to Assunguy with an emigrant by the name of Hutchings." The local Judge of Orphans, readers were reassured, then took charge of the children and placed them under the protection of "respectable families". To return the children to England, Harmer concluded, would be expensive and, as the British minister in Brazil was entirely satisfied that they were being well treated, it would be unnecessary. The eldest boy, Charles, who had a crippled hand, was said to have found a good position as a domestic servant, and only "given light work suited to his age", while the younger children were accepted more as family members.[47] Alert to the interest in Rio and beyond regarding the welfare of the Bond children, *The Anglo-Brazilian Times* reported similar information:

> After the death of the unfortunate immigrant Bond (who died of delirium tremens), the eldest daughter, who had come from England already demoralized, abandoned the other children. Under these circumstances the President of the province of Parana at once ordered the Judge of Orphans, the official guardian of minors, to take charge

of them. This the judge did, and with his consent some most respectable Brazilian families of the city of Coritiba [*sic*], voluntarily and with the proverbial Brazilian benevolence and hospitality, received into their houses and undertook the care of these orphan and deserted children.[48]

Furthermore, the newspaper stated that "the sale of immigrant children is virtually impossible in Brazil, for the feeling of the government and nation is such that an attempt at such would be viewed with great abhorrence", but the paper did not elaborate by what it meant by "*virtually impossible*".[49]

Meanwhile in Curitiba, controversy waged over the remaining English immigrants housed at Bariguy. On 15 July 1873 what the Curitiba newspaper *O Dezenove de Dezembro* reported as a "spectacular" sight occurred in Paraná's normally quiet little capital. At least thirty English residents of the lodge marched to the presidential palace in Curitiba to protest against the order that they leave immediately for Assunguy. As the official position was that immigrants should only be held at the lodge for "a reasonable period", the protestors tried to argue that, as they had by then spent months at Bariguy, the agreement had been broken and it was now reasonable to refuse the demand that they leave. The provincial authorities, under whose responsibility immigrants fell while being transferred to state colonies, were taken aback by the "obstinacy" of the delegation, considering their attitude to be "threatening public order and tranquillity".[50] Insisting that the immigrants had always been treated with perfect consideration, the authorities claimed that those at Bariguy really just wanted to remain indefinitely at the lodge and were using the most "frivolous of pretexts" to refuse resettlement in Assunguy. Acknowledging the reality that there had indeed been delays, the *Dezenove de Dezembro* – in effect the mouthpiece of the provincial government – insisted that those immigrants had been held on the Ilha das Cobras only because of the requirement to be observed for yellow fever as they had passed through Rio. While there, they had been treated well, provided with good food and, as some were in a state of penury, with blankets and clothes. The newspaper suggested that part of the problem was that the conditions immigrants enjoyed at Bariguy were the best given to any new arrivals in Brazil; food there was always of the finest quality and sufficient quantity, and health services were always available.[51]

In reply, the protestors rejected the allegations that they actually wanted to remain in Bariguy, a place where many of their fellow immigrants had died in recent months. Instead they demanded that the government should allow them a free choice of colony, one with a climate where they would be able to cultivate the kind of produce they were familiar with in England. As for enjoying plentiful and wholesome food at the lodge, they were adamant that this was simply untrue. They accepted that, under public gaze in Rio, food at the Casa de Saúde had been generally excellent, and they acknowledged that supplies at the Ilha das Cobras were also reasonable. In contrast, provisions at Bariguy were said to be appalling, with musty and mouldy beans and maggoty beef and pork, food so awful, in the words of one immigrant, "that if I had a pig at home I would not have given them to him to eat". Considering the recent experiences of the immigrants, combined with reports from Assunguy that were no doubt circulating amongst the English residents of Curitiba, it is hardly surprising that those still at Bariguy would have been weighed down by feelings of demoralization and suspicions of government.[52]

With the threat hanging over the English immigrants of all support being withdrawn if they continued to refuse to leave Bariguy, most agreed to go to Assunguy. An official report reaching Curitiba from the colony concluded that the immigrants were perfectly content with what they found there and that they accepted that they had been taken in by false information.[53] But, as with so many reports on the goings on in the state colonies, this was contradicted by the immigrants themselves. The young Clara Bond and her baby, Lucy, were amongst those were forced to leave Bariguy, in her case, according to the Cirencester emigrant Frederick Davis, having been "induced to go with [James] Hutchings up to the colony as his wife"[54] soon after his lawful wife had died at the lodge in May 1873. At Assunguy, Hutchings managed to begin to clear and plant five or so acres of the forested land that had been granted him but, in common with countless other immigrants, he and Clara soon gave up, not being given the settlement payments that were a requirement to tide new *colonos* over until the first harvests. A long-established *colono* in Assunguy said, perhaps unfairly, of the 1873 arrivals that some were entirely unsuited to deal with pioneering conditions. To him, these groups of immigrants had been sent across the Atlantic simply to increase the 'head money' that Yeats and Alsop could claim from the

Brazilian authorities. That said, he insisted that, "we are not all 'black sheep' and that there are left men of thorough respectability willing to strive or willing to 'put their shoulders to the wheel' and make houses in the backwoods."[55] Hutchings and Clara were not amongst those who persevered in Assunguy. After just a few months they left the colony, walking all the way to Antonina from where they begged a passage to Paranaguá in the hope of somehow returning to England.[56]

Chapter 7
Pioneering in South Brazil

THE VAST MAJORITY of English and Irish immigrants arriving in Brazil in the late 1860s and early 1870s were sent by the Ministry of Agriculture's colonization agency to Príncipe Dom Pedro, Cananéia and Assunguy, although a few were dispersed to other state colonies or accepted offers to settle in privately organized agricultural settlements, such as Montebello in Espírito Santos. Apart from the immigrants recruited in 1867 and 1868 by Quintino Bocaiúva, Joaquim de Almeida Portugal and George Montgomery in New York, Wednesbury and Birmingham, individual applications were also dealt with directly by the Brazilian consulates in Liverpool and London.

Also active were the partnership of Meadows and Christopher, emigration agents who were contracted by the Brazilian government to recruit one thousand British families – or an imagined five thousand individuals – per year.[1] For eighteen months in 1867 and 1868, Meadows (who claimed to have lived in Brazil for eight years) and Christopher (previously an emigration agent in England for the Cape Colony) set about recruiting English and Irish "agriculturalists" for settlement on the two hundred thousand acre portion of cleared land that had supposedly been made available to them.[2] With the assistance of a £500 grant from the Brazilian government towards their advertising expenses, placards appeared at railway stations and other public places in London and other English towns and cities, promoting Brazil as an attractive emigrant destination, a place suitable not only for the physically fit, but also for the disabled, invalids and others who would normally be regarded as wholly unsuited for the rigours of pioneering life. Would-be emigrants were invited to apply for passages subsidized by the Brazilian government and were offered one hundred acres of land at one or two shillings per acre, the fare and purchase money payable over five years.

On landing in Brazil, ten days' free board and lodging at the government's immigration hostel would be supplied, and partial employment was promised in public works schemes in an agricultural colony during the first year while awaiting the harvest of their crops. Immigrants claiming land at the state colonies agreed to pay £1 and £4 per acre as well as to reimburse the colonization agency for the construction of the hut (£3) and the clearing of the plot of land (£1). Other liabilities included having to pay part of the passage out from England. Unless an immigrant paid all this upfront (which few, if any did), an additional twenty per cent was added to the cost price of the land, with a stipulation that the debt should be cleared within six years. Whether they were aware of it or not, immigrants normally had a debt from the outset of at least £50, but some emigrants merely accepted the passage to Rio, Curitiba or elsewhere, changed their identity and disappeared.[3]

It is not altogether clear how successful this recruitment drive was, but certainly attracted both a group of young English 'gentlemen' of limited means with a taste for adventure and in search of land, a few families in reduced circumstances who were hoping to improve their situation abroad, as well as labourers, former soldiers and others recruited in London and other British cities who had no obvious agricultural experience. Some of these recruits were sent to join the Irish in Príncipe Dom Pedro, while others were dispatched to Cananéia and Assunguy.[4]

While it is probably correct that it was not in the interest of Meadows and Christopher to be at all discriminating in selecting emigrants, those who were sent to Brazil and who, for whatever reason, remained in the state colony where they were sent, exerted a vital influence on later arrivals. When the next wave of English immigrants, many associated with English agricultural labourers' unions, arrived in Brazil in 1872 and 1873, the earlier arrivals were able to share valuable experiences. The conditions that they had endured were all too typical of Brazilian state colonies regardless of the balance of nationalities, and enabled them to encourage those who were determined to remain or counsel those inclined to leave.

Colônia Cananéia

Located in the extreme south of the province of São Paulo, some three hundred kilometres from Santos, Colônia Cananéia was established by the Ministry of Agriculture in 1860. The state colony took the name of the nearby township of Cananéia, founded in 1532 and one of the first Portuguese settlements in Brazil. Described in 1865 by Ernest Buhlaw, an American surveyor, as a "friendly and inviting little town", the reality was that after over three centuries of existence Cananéia remained a complete backwater.[5] It was a small port situated on an island far from any remotely significant centre of population and with a totally undeveloped continental hinterland. "There are some towns south of Santos that make large figures upon the map," observed John Codman, an American who made a futile attempt at establishing a regular steamer route on this stretch of the coast. His failure was due to not having been able to load sufficient cargo. Instead small, locally-owned and operated sailboats carried what little freight there was – essentially rice from Iguape to Rio de Janeiro and erva maté from Paranaguá to Buenos Aires – when required. But Codman soon discovered that the print-size of place names on maps did not reflect reality, finding Cananéia, like neighbouring Iguape to the north and Paranaguá to the south, merely "wretched little villages [that] offer no inducements or conveniences for commerce [and with inhabitants] entirely indifferent to commercial enterprise".[6]

The sparse population and low level of economic development of the Cananéia region were, however, the key reasons for choosing to create a state agricultural colony there: with large tracts of publicly-owned land, immigrants would boost the economy of this remote corner of the province of São Paulo, so strengthening Cananéia as a port and population centre. The first foreign immigrants on the mainland near the towns of Cananéia and Iguape area were mainly Texans, self-imposed exiles from the former Confederate states, drawn by Brazil's apparent generous offers of land.[7] While most of these settlers soon drifted away, a few North Americans at least remained for a time, including Captain Ernest Buhlaw, a former officer in the Confederate army's Division of Engineers, who acquired land in and around Colônia Cananéia and built a sawmill to exploit the area's timber resources.[8]

Land

Colônia Cananéia was spread across an area that was located roughly twelve to sixty kilometres inland from its so-called 'port', a landing point that in reality was no more than a jetty where only canoes could moor and where there was a shed-like warehouse to store provisions or, in an emergency, newly arrived immigrants. The port was only very rarely visited other than by a canoe except by very special arrangement, while coastal steamers only irregularly called at the town of Cananéia which relied instead for its limited freight and passenger needs on small locally-owned sail boats. Despite being situated some distance from the coast, at least the colony was approached by a road of a quality that few other state colonies could match. The road was usually passable by wagons during the winter months, while only occasionally during the intense summer rains was the colony entirely cut off from the coast.[9]

The area immediately around the port was agriculturally useless, the mangrove swamp giving way to thin, scrub-covered sandy soil. Approximately four kilometres from the shore the terrain changed in character, becoming slightly undulating, the vegetation improving in quality as it entered the fertile valley of the Itapetangui. The best land in the area was considered to be that located some seven kilometres from the coast in the valley of the Rio Branco. Much of this was claimed by local figures, although many of the titles were considered to be of uncertain validity. The area demarcated as Colônia Cananéia stretched into the rich valley of the Itapetangui and it was there that the tiny administrative centre was located. Down the steeply inclined valley sides plunged fast-flowing streams that offered great potential for waterpower, a feature of the landscape that Captain Buhlaw recognized in establishing his saw-mill to exploit the timber he believed had marketing potential, although most was either *canella preta* (black cinnamon) or fig, of little use apart for making canoes. Although the precipitous valley sides were of no commercial value – even the felling of timber was impractical – a considerable proportion of the colony's land was undulating, fairly fertile and capable of yielding several seasons' crops before it became necessary to allow it to lie fallow.[10]

Families were placed far from the road and the colony centre, although because Cananéia was relatively limited in expanse they were not as isolated as in some other colonies, such as Assunguy. Nevertheless, they could be located as much as five kilometres from the road, and great effort was spent in creating *picadas* (trails) leading to their properties. Some British immigrants were fortunate enough be granted allotments on level ground, but few of these were on the side of the colony closest to the port. The problem was that most of the best-situated land was already privately owned, either by long established right of occupation or by having been purchased from the government. Efforts of more forward-looking administrators and *colonos* to purchase some of this land came to little, due to the immense difficulties involved in untangling legal ownership and resistance from Rio de Janeiro – for financial reasons and because any such purchases would have gone against the state colonies' prime purpose of settling *terras devolutas*.[11]

Produce and market

Although much of the land that fell within Colônia Cananéia's boundaries was mountainous or hilly, it was envisaged that the less precipitous slopes could be used for growing coffee and rice. Agricultural production followed the typical pattern of making a clearing no later than May, as felled trees could only be burned at this relatively dry time of year. Timing was critical: if there was a series of rainy days after land was cleared for burning, fresh undergrowth would spring rapidly to life, rendering a second clearing necessary before any planting. Clearing land involved cutting down the undergrowth and timber, the logs that had not burned left behind to rot. Seeds were then planted directly into the ashes with the aid of a simple digging stick and hoe. The first crop after cutting down the virgin forest was rarely good, but land would be left to rest for two or three years, cleared again, and a better crop would usually result. This would be planted for another three or four years, after which it would be left for a few years and allowed to recuperate. Dealing with the luxuriant growth of bushes, vines and other secondary vegetation was reckoned to require three times the amount of labour needed for similar areas in temperate Europe or the United States and, as such, families that could rely on plenty of hands had the best chance of at least short-term success. The first year was invariably hard, with *colonos* surviving largely on corn and beans, supplemented by provisions obtained by labouring on public works projects. *Colonos* who successfully overcame the crucial first year or two would next turn their attention to fencing-off pasture (horses did well, cattle poorly, in both cases put down to the quality of the grass) and in creating a kitchen garden alongside their native-style, thatched wattle and daub houses. A shallow, water-filled ditch to protect crops from being attacked by ants would surround this area, although little could be done to protect corn from the plagues of rats that frequently laid waste to an entire crop.[12] One of the few references to more exotic native fauna was a comment by Buhlaw – not perhaps the most reliable of witnesses, due to his close connections to the colonization agency, but even critics rarely dwelt on this aspect of forest living:

> The prevailing impression among foreigners that the woods of Brazil are the abode of all kinds of vermin does not seem to be correct. [...] I have seen but few snakes and other reptiles. [...] The mosquitoes are very annoying in these woods, but in the clearings near the river they do not seem to be so.[13]

Fruit of all kinds – including berries, lemons, oranges and pineapples – were said to "flourish without care or cultivation", while coffee, tobacco, rice, sugar cane, corn, beans, manioc and other produce appeared to do well, although they were not without their problems apart from those that involved animal and insect pests. The Cananéia area lacked a pronounced dry season, its climate being broadly divided between a hot and rainy summer and a warm and damp winter. While occasional floods damaged houses and ruined crops located too close to the streams, a far greater problem was that of humidity. Initially, attempts were made by North American settlers in the Cananéia–Iguape region to grow cotton, a crop many had had experience of in their southern home states, but this was quickly recognized to be impossible due to the heavy rainfall. Coffee, identified early on as potentially an important cash crop for the colony, some English *colonos* having planted several hundred specimens, was also problematic. Although coffee trees appeared to thrive, the berries often rotted before ripening or, if harvested, before they could dry in the prevailing humid conditions. Tobacco faced similar difficulties, the climate making it virtually impossible to cure and prepare for market, while even drying black beans – a basic Brazilian subsistence crop – could not be relied on. Although the climate was supportive of sugar cane, surprisingly little was grown. Only a few of the German families made any serious efforts with the crop and only one made sugar or cachaça (sugar cane spirit) mainly due to the local lack of even the most basic refining capabilities. Rice, however, was recognized as having great potential, well-suited to Cananéia's damp and hilly conditions, but production always remained limited, never rivalling that of neighbouring Iguape where it was pretty well the sole export.[14]

Regardless of finding crops appropriate to Cananéia's particular climatic and topographic conditions, the lack of a market inhibited the colony. This was despite the fact that, unlike many of the state colonies, Colônia Cananéia was connected to the coast by a good road, one that was manageable in most weather conditions. This was described by a British consular official as "although by no means a very flattering specimen of engineering, [it] is, nevertheless, one of the best dirt roads I have seen in Brazil". The road was of soil covered with a thick layer of sand and gravel, taken from the streams by which it wound, and made for "a splendid drive for either cart or carriage".[15] Furthermore, a safe sea landing-place was only some twenty kilometres from the nearest part of the colony, again a feature that few state colonies could claim. But these advantages were almost useless, given the absence of any sure means to convey agricultural surpluses to market. The *colonos* quite naturally did not care "to raise products which, if they can sell at all, must be sold at an unremunerative price, from the absence of means of shipping it and of any capitalists to purchase their produce and forward it to the excellent markets of Santos and Rio".[16] Charles Young, apparently one of the very few British immigrants with an entrepreneurial spirit, intended to create a brick works, to supply the builders of houses and the expected church and other public buildings. He felt that it would also be commercially feasible to market bricks and tiles in Rio, a project that failed due to transport limitations.[17] To say that there was no transportation was only a small exaggeration: it cost six or seven milreis to send a cartload of produce to the colony's port; the goods then had to be transferred to the island and taken the five kilometres into town. There was a monthly steamer to Santos and Rio, but freight charges were prohibitive, while in the town of Cananéia itself the only potential buyers were a few government employees and a few small traders with the power to name their own price.[18]

Population

Accounts of the local population immediately prior to the arrival of the first foreign immigrants in Cananéia in the mid-1860s are scarce. Captain Buhlaw, however, described the native population of the area as being of "thrifty appearance" and with a "friendly disposition towards strangers". They were "public-spirited" and a "most intelligent, hospitable, Patriotic and public-spirited set of men and can boast in addition a host of Lovely Women, who try to make the Strangers Welcome".[19]

The local population was certainly very small in number and if there were contacts with land-seeking immigrants they do not seem to have been a cause of tension. Apart from a few North Americans linked to the Texan colony centred more on Iguape, only in 1868 were foreign immigrants – Germans, Swiss, English and Irish – being directed to Cananéia in significant numbers. The first arrivals from England included the contingent of passengers from Birmingham who had travelled to Rio with the *Florence Chipman*, as well as a number of immigrants recruited in London. Of these, none had agricultural experience.

Although the number of immigrants from England who went ashore at the colony's port totalled several hundred, only a small proportion of these actually made an effort to settle. The 1872–3 arrivals were diverse in background, including Gloucestershire and other West Country town folk and agricultural labourers, as well as more immigrants from England's industrial heartlands. Typical of these was Joseph Kilchen, formerly a toolmaker at a small arms manufacturer near Birmingham, who reported having been attracted by glowing accounts of Brazil, a country where working men "could get on well in farming and work would be given on roads and bridges". James Davies, from Birmingham, was similarly persuaded to leave his job as sawyer to begin a new life with his family in Brazil.[20] Yet another emigrant was Edward Lyons, also from Birmingham, who at the age of 52, was accompanied to Brazil by his wife and two children. Claiming to be unaware that he was going to be sent to an agricultural colony, Lyons later explained that he had thought that he was going to São Paulo to work as a lamp maker, his only skill. But the Brazilian consul in Liverpool advanced Lyons and his family the passage money, less £5, apparently on the basis that he was an agricultural worker. Whether Lyons was aware of this is unknown, but being illiterate he would have been placing his 'mark' on a contract that he could not read. After a five-week stay at the Casa de Saúde in Rio, Lyons was sent to Cananéia.[21]

By late 1874 there were 124 British inhabitants in Colônia Cananéia, a population made up of 23 families and three single men. Despite marriages between British, Swiss and German settlers being commonplace from the colony's early days – bonds facilitated by the fact that most of the inhabitants of all nationalities were Catholic – the English and Irish immigrants were often viewed with contempt and regarded by the colonial administrators as "vagabonds", unfairly foisted onto the colony by officials in Rio. But being criticized as "sluggards" contributed to a sense of solidarity amongst the English *colonos* who rallied in their own defence: as they declared to the British minister in Rio in 1870 "[w]e are all British subjects and we believe of sufficient intelligence as working men, to know that we have to get a living by hard toil".[22] Carlos Brent Cenci, an American sent in 1874 by the British consulate in Rio to investigate conditions in Cananéia, was one of those critics of the quality of the British families, reporting that many "would prove a pest in any country, in any community, under any circumstances". According to Cenci, most of these chose to survive on whatever work was made available by the colony's administration, spending "most of their earnings for strong drink, and much of their time in idleness and creating disturbances in the Colony". Even so, Cenci considered such individuals to be in a minority, with most of the immigrants being viewed as hardworking, industrious and orderly. That said, he agreed that the German *colonos*, who had arrived at much the same time as the first British, had achieved more, being "generally more industrious, [...] far more frugal in their manner of living, and as a rule they drink less cacháça".[23]

Administration

Of the many hundred British immigrants who were sent from Rio de Janeiro to Cananéia, only a small proportion of them were ever issued with allotments. Despite the promise of several acres each of cleared land ready to plant with a shelter, few newly-arrived immigrants were so fortunate, most being forced to camp out for months on end. This would be reasonable for a self-sufficient pioneer settler, but the great majority of British immigrants who had agreed to join a Brazilian state colony did so because they had no independent resources and were attracted by the generous settlement package they had been offered. State colonies were renowned for being under-funded and found it difficult to hold on to personnel who sought every opportunity to spend time in the Brazilian capital rather than, as in Cananéia, in the storeroom that served as the administration's offices, surrounded by old harnesses, blacksmith's and other tools. Cananéia's situation was probably similar to that in the other state colonies, invariably placed in remote parts of the country, where it was difficult to retain a full complement of personnel, even if they were not particularly capable individuals. For example, when Cenci visited Cananéia in 1874, he found that while the colony's director, Dr F.C. Figueiredo, was generally respected, he was frequently absent, his post taken by his assistant, a Senhor Vanderhoff, who was "universally despised, and, in my opinion, justly so". But as the immigrants were utterly dependent on the colony's administrators, they rarely risked voicing their complaints directly.[24] Within the isolation of a state colony, a director could take on the role of petty tyrant, there being no realistic means of appeal by *colonos* apart from complaining to the British legation in Rio in the hope that they would take up the issues with the Ministry of Agriculture.[25] Director Barrata was considered one of the worst offenders, accused of forcing *colonos* to buy inferior quality goods on credit at a higher cost from a particular store with which he was associated, and of driving *colonos* away so as to claim for himself the land that they had improved – a common predicament for pioneer settlers whether in nineteenth-century Texas or twentieth-century Amazonia:[26]

> The Director is very tyrannical. He took a fancy to certain cattle belonging to Joseph Gibbs, an old man of 60, very industrious and generally respected in the Colony, and because he would not sell them to him, he began to annoy and oppress him, took away the land that had been given him, forbad him to use it for cattle and appropriated it for himself, he also restricted the time for doing work given to Gibbs by the Commissioner of the Imperial Government to such a degree, that it was impossible for him or anyone else in the colony to do it in time.
>
> When William Watts applied to the Director for work or assistance, having been shifted twice, through mistake in the land given to him, and losing one house built on the wrong land, the Director replied that he could give neither, and when Watts observed that in that case he would be starved to death together with his family, the Director replied, 'Die then and I shall be rid of you'.[27]

Apart from the distribution of allotments, it was the granting of work that made a colony's director so important in the life of an immigrant. *Colonos* quickly came to realize that, without paid work, it was impossible to support themselves from their lands alone. Employment was essential, even for an established *colono* with many acres of cleared land at their disposal, and it was quite simply impossible for a newly arrived immigrant to survive without paid work.[28] Occasionally a director went out of his way to help an immigrant find specialist employment – as Mr Figueirdo attempted to find work in his profession in Rio for Edward Lyons, the Birmingham lamp maker – but it was road-maintenance or similar work that a director might have in his power to hand out.[29] The immigrants insisted that government was contractually obliged to provide them with fifteen days of work per month, a claim that was to be a constant source of dispute.[30] As early as 1868, a petition was circulated protesting the lack of government payments. "We have no wish to live on the resources of the Government," it addressed the minister of agriculture, "but we beg to say that the Colony cannot as yet sustain itself and unless we get assistance from Rio we must fail; but with assistance we are confident it must and will eventually prosper."[31] On occasion the British legation in Rio intervened on behalf of the immigrants. On one such, the minister of agriculture stated that the immigrants were not being paid as they had not been issued with a valid contract locally, to which George Buckley Mathew tersely responded, "surely His Excellency cannot mean that in doing work with their Director, these poor people should require a formal contract, instead of trusting to that functionary's verbal agreement?"[32]

When payments were forthcoming, they took the form of paper notes issued by the colony's administration, only redeemable at its store and then at below face value for goods sold at over-inflated prices. Immigrants usually had to contend with few supplies, receiving no fresh provisions for weeks at a stretch and reaching a point where they claimed they were in a "state bordering on starvation".[33]

Social life

Those immigrants who persisted through the first season and were prepared to live on a mix of subsistence agriculture, supplemented by occasional periods of paid employment, had to contend with isolation from all that they had previously known. Admittedly, there was a cricket club, which met on every pleasant Saturday afternoon, initiation fees going to support a lending library that, at the end of 1874, boasted 75 volumes in its collection.[34] Other important points of contact were the frequent public meetings held in the colony to protest against conditions, demand payment for work done, provide assistance to those who wanted to leave and to ask the British diplomatic and consular officials in Rio to make representations on their behalf. Although these gatherings rarely secured any directly positive results, they brought otherwise isolated *colonos* together, breaking their isolation to act in a common purpose. A frequent cause for complaint, despite the colonization agency's regulations, was the fact that there was neither a Protestant nor a Catholic (the faith to which most of both the British and German immigrants adhered) church in the colony. In addition there was only a school for small children, attended in 1874 by no more than three or four of the British. Calling for an English school to be established and for a priest to be sent to the colony, British residents explained to George Buckley Mathew that "[o]ur children are growing up in a state of ignorance with regards to Religious instruction, or learning of any kind".[35]

The relationship between alcohol consumption and frontier living is complex. For many of the immigrants, alcohol dependency had already been in England a way of life in which drink was part of wages, sometimes credited with health-imbuing properties. Isolation and other factors are likely to have made many settlers – regardless of national origins – consume more alcohol than in their home countries and, whereas beer or cider would have been the most common beverages in England, in Brazil cachaça was the sole form of alcohol available. Certainly the British *colonos* consumed a great deal of alcohol, but Britain's consul in Santos felt that what they had experienced was "enough to turn better men to drink".[36] In his detailed 1874 report on Cananéia, Carlos Cenci was rather less understanding, believing that too many of the settlers were reliant on drink to a destructive degree. Michael McCornville, for example, an elderly former soldier and pensioner, two years in the colony, was said to be "utterly worthless from strong drink", Mrs McCornville, who was a "quiet, hardworking woman", had the misfortune to be "often turned out of the house at night by her husband's "drunken carousals". William Watts, who had managed to clear some fifteen acres of land in two years, was dismissed as "neither very industrious or sober".[37]

Others, in the opinion of Cenci, could cope better with their habit, although he remained judgemental, hardly wavering from a view that intemperance at the least conspired against long-term success. Despite being "a little too fond of cachaça", Cornelius Murphy was considered to be "a very industrious man" who had succeeded in clearing fifty acres and was "very well established" in a very good location. William Johnson, who only arrived in 1873 but was seemingly "trying to get on", had ten acres cleared, "and although he does not neglect his business, still drinks a great deal". Of James Davies, who had cleared eighty acres since arriving in the forests in 1868, Cenci commented that he "drinks very hard, and were it not for his wife he would be utterly worthless in the Colony, and even now he is often in drunken broils". Joseph Gibbs and his son, while it was acknowledged that both were hard working and quiet, were said to drink a great deal of cachaça; for this Mrs Gibbs and her daughter were both "deserving of sympathy".[38]

But even those who lived blameless lives appeared to fare no better than those who were reduced to drink. John Barton, Elijah Fulcher and Michael Murray were amongst the men who were praised for being "sober", "quiet", "hardworking" and "industrious" but whom, nevertheless, were anxious to leave. Cenci was only able to identify a handful of families who expressed satisfaction with conditions, or at least maintained some hope for improvement. Joseph Jackson who lived with his wife and their young child, was content, thanks to success as the colony's baker and in keeping a small shop. John Nichols, his wife and seven children, and Richard Monks and his family were discouraged by the present condition of the colony but expressed a desire to carry on living there.[39]

Colônia Assunguy

When the province of Paraná was created to the south of São Paulo in 1853, its overwhelmingly rural population was essentially made up of two social classes – the *coronéis*, land owners with attendants to enforce their authority, and share-croppers and squatters who allied themselves with a *coronel*, or occupied a small piece of unclaimed land where they practised virtually subsistence farming. The relationship between these two groups was complex in a society composed of overlapping lineage or extended families which included legally constituted members, illegitimate branches and clients, all of them under the patronage of a *coronel*. Although there was not always any obvious difference in wealth between a *coronel* and his dependants, he could wield an authority that resembled a master/slave relationship. The territorial claims of a *coronel* were largely based on the strength of kinship ties and alliances with other *coronéis* rather than on the strength of any legal title. But even when title was held (often the result of the Portuguese colonial *sesmarias* – land grants – awarded to soldiers on completion of military service) the land was of little value without labour to exploit it.[40]

Rural Paraná was integrated into the wider provincial economy through trade with small population centres in the interior such as Curitiba, Castro and Ponta Grossa, and with São Paulo, Rio de Janeiro, Minas Gerais, Argentina and Uruguay using the north–south cattle and mule trails, and the ports of Paranaguá and, to a lesser extent, Antonina. The province of Paraná was best known for the production and export of erva mate, a wild holly-like shrub from which leaves were collected and toasted for a tea-like beverage popular in the south of Brazil, Argentina and Uruguay. Pigs were also raised in the interior, in part for local consumption but mainly for conversion into easily transportable forms such as lard, bacon and skins, for sale both within and outside the province. The local method of pig production was well suited to benefit from an abundance of land over labour. During the winter months the animals could be virtually ignored, released into the forest to scavenge for food amongst the wild fruit and pine kernels that had been brought to the ground by the heavy rains. At this stage all that had to be fed to the pigs was salt. When they had reached a few dozen kilos in weight, the pigs were passed to *safristas*, slightly wealthier *posseiros* for fattening. In winter *safristas*, sometimes employing contract-workers, cleared some of the forest by setting fire to it, maize was planted in the ashes and, as the plants reached maturity, the pigs were released into the fields. When sufficiently fattened, the pigs were either taken by hoof to market or slaughtered for bacon, lard and hides.[41]

It was at the distribution stage that the *fazendeiros* – large land-owners – or those closely linked with them were at a particular advantage in the marketing of erva-maté and pork products as well as small surpluses of agricultural staples such as corn and beans, a condition that the English surveyor Thomas Bigg-Wither observed during his travels in Paraná in the early 1870s:

> From want of proper roads, the whole traffic with the prairies has to be carried on by means of pack mules alone. In order that this may

be done profitably, it is necessary that each large *fazendeiro* or farmer shall have a sufficient pasturage on which to keep a troop of these animals. Hence it is that small landed proprietors are heavily weighted in the race for prosperity; because, not being able to keep a mule troop, from the want of sufficient extent of pasturage, they are forced to sell their crops *in situ* to the larger *fazendeiros* at their own price.[42]

Not being part of a powerful social network, with little in the way of government support and few resources of their own, the English and other immigrants in isolated state colonies such as Assunguy found themselves virtually marooned, totally dependent for marketing their surplus produce on what was charged by mule owners.

Immigrant settlement

When the province of Paraná was created from São Paulo in 1853, its population totalled just over sixty thousand.[43] Both the central and provincial government saw the arrival of immigrants as an essential step to broaden Paraná's limited economy. Government and private initiatives resulted in the establishment in the 1850s of several agricultural colonies on the province's small stretch of sub-tropical coastline and in its forested interior. These colonies, isolated and poorly financed, failed to prosper and the population of the province grew only slowly. Most of the immigrants drawn to Paraná settled in and around Curitiba, the province's capital and main commercial centre. But even in 1870, Curitiba had only about four thousand inhabitants, with perhaps as many more again within a radius of fifteen kilometres. Of the rural population, about 2,500 were German, while most of the others were either Polish or French.[44]

Located about one hundred kilometres north of Curitiba, the state colony of Assunguy was founded on 21 October 1858. Of all the government-maintained and private agricultural colonies, it was Assunguy that was to attract the largest number of English and Irish settlers, with over a thousand either settling or passing through by September 1874.[45] On the face of it Assunguy appeared a rather more sensible location for a colony based on northern European immigrants than either lowland Cananéia or Príncipe Dom Pedro, which was constantly at risk from floods. Assunguy was situated three hundred metres above sea level, with a climate that was considered perfectly healthy although it was very hot in the summer, during which thunderstorms and torrential downpours of rain were common occurrences.[46]

The colony's first settlers were eight German families who arrived in 1860, followed soon after by a few others. In 1862 the colony's total population was 85, and two years later it reached 197, composed of fifty Germans, three Swiss, one French and the rest Brazilian, although many – if not most – of the people in this category would have been born in Brazil of foreign parents.[47] During the remainder of the 1860s, Germans continued to arrive in Assunguy, mostly transferring from the comparatively successful Blumenau and Dona Francisca (Joinville) colonies where land had become increasingly expensive and in short supply. This movement was the cause of tension between government officials, with Paraná said to be unfairly inducing Germans to abandon Santa Catarina, a serious accusation given that northern European immigrants were being so assiduously pursued.[48] The Paraná government forcefully denied that Germans were being deliberately lured into the province and declared that it was simply impossible to impede their daily arrival, despite the considerable natural hurdles, not least the dreadful roads that linked the two provinces.[49] Whether there was, or was not, official connivance, there were certainly those in Assunguy who were keen to attract Germans, offering the familiar reason that "they work so hard in whatever industry they enter [and] also because in general they connect most easily to national customs".[50] British – and even less the French – were certainly not the first choice, with a Brazilian resident of the colony pointing out that "[e]veryone who knows American history understands that it was not mainly the English and Irish who populated and cultivated the wastelands lands of North America, but it was the Germans who transformed them into productive fields by means of the hoe".[51] More to the point, Germans from Santa Catarina were already familiar with pioneering in the forests of southern Brazil and would not be as intimidated by local conditions as were likely to be immigrants newly arrived from Europe.

But for a while it was French who rivalled both the German and British immigrants as the most noticeable new presence in Paraná. French hotels and shopkeepers were a common sight in Brazilian and other South American cities, but French immigrants in Paraná were, unusually, rather more primarily attracted by farming opportunities.

Coinciding with the arrival in late 1868 and early 1869 of Assunguy's first contingent of English immigrants, were several French groups who had been assembled by Brazil's vice-consul in Marseilles. Their origins are not entirely clear, but it appears that they were farmers from Oran, in the French colony of Algeria, who fled to Marseilles after locusts and drought destroyed their entire crops. Many of these immigrants were given land in the newly-established Colônia Argelina, situated on the periphery of Curitiba, but when all the available land there had been taken up, the others – against a background of considerable protest – were ordered to go to Assunguy or else have all government aid withdrawn.[52] There are few indications of how these immigrants fared, but certainly the French presence remained a strong one for years. British and Brazilian sources rarely mention them although they made up at least thirteen percent of Assunguy's 1874 population.[53] Thomas Bigg-Wither, the English explorer and surveyor, left an unflattering account based on conversations with one of the colony's store-owners who kept a large ledger listing the names and debts of every male colonist. He dismissed the French as being "'ladrões todos'" – thieves, all of whom were "utterly worthless and despicable", never paying a single debt, contrasting them with the English who were "'boms rapazes'" – good lads who got drunk and fought, but seldom forgot their debts when they had money to pay them.[54]

After arriving in Rio de Janeiro from Europe, Assunguy-bound immigrants were first sent to Paranaguá, Paraná's principal port, located two days south of the capital by steamer. After a period in quarantine at the Ilha das Cobras, they were transferred by boat to Antonina, a small port on the western edge of the Bay of Paranaguá, from where the road inland began its winding ascent up the forbidding escarpment. The route was still little more than a trail when the first English immigrants made their way up to Curitiba, but in 1872 a well-maintained macadamized road was completed, allowing wagons to cover the distance. From Antonina, the road carried on up the Serra do Mar mountain chain, crossing at its highest point of over one thousand metres above sea level before descending again by 120 metres to reach Curitiba. Germans had almost exclusive possession of the wagon traffic between Antonina and Curitiba. The road's tolls were considered to be extremely high, with every wagon having to pay the equivalent of about £1 6s to cover the route, making it extremely expensive to import goods and often hardly

A panic upon a bridge. (British Library)

worthwhile to send to market produce from the interior.[55]

Had the 'road' that linked Curitiba with Assunguy been a good one, the distance could have been accomplished on horseback in one day with the aid of a second mount, or by wagon in two and a half days. But for most of its distance, the 'road' was no more than an extremely poor mule track, with the journey usually taking four days to accomplish, and in poor conditions six days or more. The more fortunate men who were travelling between Curitiba and Assunguy would ride on horseback, but most would walk. Women and children would usually be carried in boxes, one on either side of a mule, a box typically capable of holding two small children.[56]

Most of the streams and rivers that the Assunguy trail crossed were without bridges and in rainy weather the larger ones became impassable, suspending traffic for hours or days on end. While there were some stretches of the road along which carriages could pass, in reality the route's practicability was only that of its narrowest point and worst surface. The first 25 kilometres from Curitiba was through open plateau, good during dry weather but in wet becoming a total quagmire. Instead of the grassy plain that surrounded Curitiba, the whole area north of this point was of broken hills, mostly a mix of *capoeira* (secondary vegetation) and *roças* (forest clearings for agriculture). Paths that led to *roças* were numerous, and easy for a traveller to get lost in while trekking to Assunguy. Soon afterwards the road entered mountainous terrain, steep ravines and narrow valleys covered with dense Atlantic rain forest, with the track – when first cut, about two metres in width – diminishing in places to bare footholds for a mule. The route here was slow and often dangerous, especially in the rains; carcasses of mules that were bogged down in deep mud or had fallen with a load at a narrow pass marking the dangers. Along especially muddy stretches where mud could reach a mule's saddle girth, the animals would be unpacked and left to find their own way to a dryer surface, their handlers carrying the goods being transported on their shoulders. The track hugged the nearly perpendicular mountainside, the route barely wide enough for a single mule – let alone two – to pass, while torrential rain made the red soil as slippery as soap, the pack animals constantly losing their footing.[57] As described by Bigg-Wither:

> We had enough to do in attending to our mules, which were performing erratic slides on the slippery surface of the steep paths, and threatening momentarily to bring one or other of us to grief, to think of anything else. [....] My note-book records the fact that on this day my mule tumbled down five times, whereas once or twice per diem was generally the average on bad forest tracks. It must not be thought that when a mule tumbles, its rider necessarily tumbles too. On the contrary, he merely spreads his legs apart to avoid having them crushed – for the fall is more often caused by a side slip, than a direct forward tumble – and waits in the saddle till the animal gets up again. Sometimes, however, the shaking that he receives is very severe, for the mule, if it be a plucky one, will make the most frantic efforts to regain his footing, which efforts frequently render matters worse, and culminate in a rib-breaking fall which jars the rider's backbone to its very core.[58]

The slopes were so steep that riders of mules and horses needed to constantly to dismount and re-adjust their saddles. When two files of animals met at narrow spots, some would inevitably fall into the ravine below, perhaps a hundred metres or more, resulting in death to the mule and great difficulty or impossibility of recovering a load that may have rolled into a river. The usual halting place was Votuverava, a hamlet established at the beginning of the century and which consisted of about eight houses and a chapel. From there, for some unknown reason, the engineers had selected one of the most difficult choices of terrain to create several kilometres of carriage road at considerable expense, only to suddenly suspend it at a narrow pass. Construction work on the road was constantly being abandoned and then re-started according to the whims of Curitiba, "one Engineer undoing what another began", as one exacerbated settler remarked, while Bigg-Wither, the English surveyor, calculated that at the rate that work was proceeding, the road would take two hundred years to complete.[59]

Curiously, though, observers believed that constructing a wagon road would not have been especially difficult, as at most points along the route there was decomposed granite near the surface or rocky outcrops no more than a metre high. About a mile outside the village was a piece of macadamized road a few hundred yards long, "the miserable result of many long years of expectation for the promised highway to Curitiba".[60] When it was decided to start road works seriously, the provincial authorities began at a point equidistant from the two places it was intended to connect, whereas it would have been far more sensible to have started in Assunguy where *colonos* were always eagerly hoping for paid work. Despite the obvious lack of progress on the road, the authorities always held out hope that communications would improve, visiting dignitaries being assured in August 1873 that a trail to Antonina was also being planned, enabling goods carried to or from Assunguy to by-pass the capital.[61]

Administration

Colônia Assunguy's administrative centre was attractively situated, located near the mouth of the small Ponta Grossa River, a tributary of the Ribeira River. Here, the rich valley suddenly widened to a breadth of over one kilometre, providing a large piece of land that had been appropriated for the creation of a village. Most of the village buildings were ringed around a one and a half acre open quadrangle, known as the Pateo da Colônia, about half of which consisted of a low mound of earth and a swamp. Despite the suggestion of the colony's directorate, settlers and British officials alike, that the mound should be levelled into the swamp to improve local health, decades passed before this was carried out. On one side of the Pateo were single-storey, whitewashed buildings, the most prominent of which, marked out by palisades painted in the green and yellow colours of Brazil, were the private residence of the director and the administrative offices of the colony, where as many as 23 government officials were employed. Alongside these buildings were the cookhouse, a hostel – in reality closer to a barn – for newly arrived colonists, stores, a medical dispensary and, a little distance on the same line, just beyond the quadrangle, a small Roman Catholic church. To the right of these buildings was the small cottage hospital and a tile-kiln; on the left, a waterwheel, sawmill, and eight or so small houses; on the fourth side were a number of privately-owned buildings including a French-run hotel where food was also served, bakeries, and three or four surprisingly well-stocked general stores that boasted that they supplied everything "from a Brummagem tin kettle to Manchester cotton goods". Beyond this area were about a dozen other small houses inhabited by government-employed land surveyors, artisans and mainly German shopkeepers, while orange groves punctuated the entire village. The cemetery was some distance behind the government houses, with the Protestant separated from the Catholic burial-ground by a fence, with a large wooden cross placed in front of each, a demarcation that was not sufficient for some religious purists.[62]

Agriculture

The area near Assunguy's administrative centre was especially suited to agriculture, although only the Patio da Colônia and the area immediately around it would have been suitable for ploughing. But, apart from a few German and French *colonos*, immigrants were assigned land at some considerable distance away from the administrative centre as most of the colony's prime land had long before been ceded to influential provincial figures. Their properties commenced virtually at the door of Assunguy's administrative buildings: for example, eight or nine sections (a section being three hundred acres) belonged to non-resident Senhor Miro, eight sections belonged to a Senhor Rocha, also a non-resident, and a further eight sections were owned by Severo Correa who lived in the colony but, like Miro and Rocha, had not put any of his land under cultivation. A consequence was that newly arrived immigrants were located as much as thirty kilometres, or even further, from the village, in some cases where several streams flowed which, when swollen by heavy rains, could be impassable for many days at a stretch.[63]

Approaching Assunguy from Curitiba, the first farm allotments were found approximately twenty kilometres south of the colony. Other English immigrants were placed to the north at various intervals, on rich, but steep land alongside the Ribeira and Turvo rivers, at least a half-day's journey on foot from the colony's administrative and commercial nucleus. A third line of *colonos* had been allotted sections between twelve and twenty kilometres from the centre on the Jaguatericu River.[64]

The size of properties allotted to newly arrived immigrants varied considerably, generally being half a section (150 acres) for most of the 1868 arrivals but just 35 or 75 acres for the 1872 and 1873 arrivals. They were sold regardless of quality for the equivalent of two shillings an acre, repayable over a period of seven years. It was particularly difficult for the 1872 and 1873 immigrants to secure a property. In theory an immigrant could select his own land, but in practice he would be sent many kilometres from the colony's centre in total ignorance of what was available and what features to consider in selecting a plot. Typically, the would-be *colono* would then be repeatedly told that the chosen land was not available, either because it was owned by, or was being assigned to, someone else. Months would elapse before a newcomer was issued with land, during which time a large debt had accumulated for supplies that they found difficult to eat. Complete confusion seemed to reign in Assunguy's land survey department with lots often only measured after occupation, and frequently families found themselves to be living on a property that did not belong to them. For example, Gloucestershire emigrant Henry Robins built two houses, and partially cleared and commenced cultivating two separate allotments, before being required to transfer to a new section of land.[65]

Colonos' houses – remaining mere *ranchos*, or huts, for many years after the arrival of the immigrants – were invariably built of palm, constructed in the local style with wattle and daub, typically situated on a clearing about 35 metres by 20 metres. In October, beyond the clearing on which the rancho had been erected, a *colono* would fell seven or so acres of forest which would be left to dry for a few weeks before being set alight and reduced to ash and planted at intervals of just over a metre for maize and a little more than that for beans, dropping into each hole three or four seeds. In January or February weeds were cleared away and in May these crops would be ready for harvest with a one hundred-fold return in a good year. The next year the process would be repeated, but with fertility soon reduced, a new piece of land generally then had to be cleared: that which had been sown the previous year being left to lie fallow for four to six years before the wild growth was again ready to be cleared and burnt. But where steep land – generally rising at angles of between 35 and 75 degrees from the level of the mule paths – was cleared, the heavy rains rapidly carried away the greater part of the rich but scanty alluvium into the rivers. A settler on this soil, therefore, could be in a worse position than one cultivating the poorer quality, but clear and level, soil of the tableland around Curitiba that offered the possibility of improving year-on-year with the limited available manure. Due to the labour involved, the fact that land could only be planted for a few seasons before it needed a period of rest and that the contours of land such as that granted to most of the immigrants were such that a plough could not be used, clearings were never grubbed, and only superficially cleared of forest before burning.[66]

The testimony of William Ryan, an immigrant from Lincolnshire who went with his wife and six children to Assunguy in 1873, describes experiences typical of those who took up land in the colony. After cutting a path to link his property with the nearest river, Ryan cleared three acres of land, felling trees and leaving brush measuring over two metres high and dense green thicket. When cut and burned, the clearing had "the appearance of a Church yard with Rock standing up high". According to Ryan, the soil was "rich and good [but] when it came exposed to the hot sun it was [so] dry it would run through a sieve". It did not prove a success with the potatoes that Ryan planted with great difficulty – the tubers failed to form although the green tops grew in abundance. Mules entered the field and consumed most of his first crop of beans, but he was able to harvest some corn – where the soil was thick the ears developed well, but elsewhere the plants withered. He then employed a fellow immigrant to fence in his clearing, only to find that mules were able to push their way back into the field and eat what was left of his crops.[67]

The quality and gradients of land, and the distance from the village of the allotments, were not the immigrants' only immediate obstacles. Many properties could only be reached by crossing rivers and were also a considerable distance from the main access trails. Access to the

The first stage of the process of clearing the forest for settlement. (British Library)

sections on the Jaguatericu River, for example, could only be managed by first crossing the deep and fast flowing Ponta Grossa River which during the rainy season was often completely impassable for days on end when the waters would swell and the currents were at their most treacherous. When bridges were eventually constructed across rivers they were usually positioned with no apparent thought. For example, *colonos* complained that it took over a year for a bridge to be built, at great expense, across the Jaguatirca River, but its location effectively made it useless. Known to Assunguy's English residents as "the Brazilian folly", there was no trail leading to this bridge as the Brazilian owner of the land on one side of the river denied access. As a result, the forty or so Brazilian, English and French families living on the opposite side of the river who might have benefited from the bridge had to continue to cross the river chest-deep in water at other spots, and during the rainy season were unable to so for long periods of time.[68] The trails leading to individual allotments were routed with no apparent purpose: for example, Frederick Tigar's farm was accessed by a track that wound about "like a snake up hill and down dale, twice the distance it ought be" and where after heavy rains the user would be "knee deep in mud".[69]

The *colonos* calendar revolved around clearing, burning and planting their land, typically requiring up to twelve acres of virgin forest or twenty acres of *capoeira* for a family to cultivate each year.[70] Corn was planted in November and harvested in May and June; beans planted in July and August for harvesting in September and October; tobacco planted in November for harvesting in April; sugar cane planted in February and March for harvesting throughout the year.[71] Of these products it was corn and beans that invariably were the first seeds that a newly installed *colono* would plant, but the immigrants – who either had no previous farming experience or had laboured in the altogether different agricultural environment of England – experimented with varying success with a wide range of other crops.

Brazilian agents in England had led prospective emigrants to believe that Assunguy would be equally suited to the production of temperate and tropical crops. Apart from corn and beans, manioc, potatoes, sugar cane, coffee, tobacco, grapes, oranges, bananas and arrowroot were the main crops produced. But even success with these staples varied from property to property. Some *colonos*, for example, were able to produce good-quality sugar cane, while others could not, and coffee did reasonably on the frost-free locations alongside rivers but not on properties on higher ground.[72] Packages of up to five pounds in weight (approximately two kilos) could be sent without charge by friends in England, but were always held up in Rio – and usually also in Paranaguá and Curitiba – taking over six months (and up to two years) to reach their destination, if they arrived at all. Frederick Tigar was one of those who received seeds from England, although he recalled the frustration of receiving a case containing rotten currant and gooseberry seeds. While the seeds did not have even a chance to develop into the reassuringly familiar bushes that the sender would have hoped for, in all probability the gift would have been entirely inappropriate for Assunguy's subtropical conditions.[73] Occasionally cuttings or seeds were delivered to Assunguy by the colonization agency in Rio, but without a local system for delivering agricultural advice or support, these efforts went unappreciated. For example, olive seeds were distributed but the *colonos* refused to plant them on the grounds that far too much time was required before they might develop: by necessity, their horizons were unable to extend beyond the next planting season. They were unwilling to risk spending effort on experimenting with a crop of which in any case they had no experience and a product of which all were entirely ignorant. Cotton, however, was one potential cash crop that was attempted, but the plant performed poorly due frost on higher ground and insufficient flat land, best suited to extensive agricultural production, alongside the rivers.[74]

But production of even the crops that grew successfully – whether corn and beans or more obvious cash crops such as sugar, coffee, arrowroot and tobacco – was small scale due to transport problems. Virtually none of the *colonos* kept mules due to a small producer's immense difficulty in creating and maintaining pasture and they would be forced to pay a prohibitively high price for the hire of mules if agricultural produce was ever taken to market in Curitiba. A mule carrying a 240-pound cargo of corn the one hundred kilometres would cost five milreis (ten shillings) when the value at market amounted to precisely the same.[75] Given the work involved in producing the corn and also the time on the road, taking corn to sell would incur a loss. The only product that could sometimes turn a profit was bacon – most (but not all) *colonos* kept pigs, useful consumers of surplus grain – but finding mule owners, often linked to their productive competitors, to carry the meat to market could be problematic.[76] Occasionally people were able to sell their surplus to Assunguy traders and storekeepers, but at extremely low rates that only went to defray expenses, covering little more than transport costs from farm to village. Compounding this lack of an effective market was the fact that, instead of purchasing produce from Assunguy's *colonos*, the colonial officials usually ordered bacon, beans and other basic supplies from Curitiba, at one time from a store said to belong to the brother-in-law of the provincial vice-president, paying twice the price that local produce would have cost.[77]

7. Pioneering in South Brazil

'The Grange': the Tamplin family's farm. (Tigar family)

Employment and debts

On arrival in Assunguy most of the *colonos* received the expected head and seed money to purchase basics such as potatoes, manioc, beans and corn for planting, calico cloth, gunpowder and shot. This was meant to tide them over until their farms entered production, but merely increased the debt the immigrants already owed the colonization agency.[78]

Immigrants recruited in 1872–3 by Brazilian agents and consulates in England were promised fifteen days work per month in the state colonies on road or similar projects for their first half year or year.[79] In reality, such work was only occasionally available – and even then, pay was very often withheld as the colonies' treasuries were more often than not empty. But paid work two or three times a week was a virtual necessity if the immigrants were to be able to stock their farms and repay the government for the debt accumulated since leaving Europe. The English immigrants also claimed discrimination, in 1875, for example, pointing out that only four of them had been given work constructing Assunguy's new church, schools, hospital and parsonage. Instead, Augusto Kaisner, a German living in Curitiba who had most of the construction contracts, had brought most of his workforce with him. Assunguy's engineers and directors were also accused of favouritism regarding road maintenance work, the English believing that there was a tendency to favour German *colonos*.[80]

Instead of receiving cash payments for work, *colonos* usually at best received a letter of credit. To redeem these could be impossible and out of desperation *colonos* generally sold them on to traders for as little as one tenth of their face value. When payments were not issued for work, or if a worker insisted on claiming the full value of a promissory note, the *colonos* were forced to apply to Curitiba, but officials there could only respond to a *requerimento*, a formal written request on stamped paper. Unable to write Portuguese, the claimants had no option but to pay an amanuensis one milreis for this work (plus the cost of the stamp), money that they could not afford. Nevertheless, the demand was so great that Sandon, a French shopkeeper in Assunguy, on one occasion purchased fourteen milreis worth of stamps that were all sold to validate *requerimentos*. Some *colonos* addressed several *requerimentos* without receiving a reply and many walked several times between Assunguy and Curitiba, in vain seeking payments of small sums of money that officials readily admitted they were owed.[81] But even when payments were eventually granted, the effort involved was sometimes barely considered worthwhile for what were often trifling amounts, even for cash-strapped immigrants. As Frederick Tigar reported:

> During the busiest time (roça time, cutting and burning timber so as land may be made ready to receive corn) we were sent to Curitiba, some 20 of us, to receive money the Director owed us for two years, paying our own expenses, in some cases [....] more than the amount to receive, this 'by order of the President' and was considered a *favour* that he had granted the money to be paid.[82]

With their budgets always overstretched, Assunguy's directors were forever complaining that the colony's treasury was entirely empty and found it increasingly difficult to face the *colonos*. Whenever a director returned to the colony from a visit to Curitiba, his creditors – many of whom had travelled from outlying areas – would besiege him, although they almost always returned home disappointed. "The consequence of this non-payment", wrote Tigar, "is that the director must frame excuses and descend to the merest subterfuges to get rid of the poor colonists to whom the Government owes a few paltry milreis." On the rare occasions that some money was handed out, it was immediately passed on to the storekeepers who had provided credit for provisions.[83] By 1876 the demands from the *colonos* for payments had become so constant that, according to Tigar, the director started "shutting himself up so no one can have an audience with him".[84]

Population and social life

Assunguy's inhabitants were widely dispersed, with the administrative centre of the colony being the only village. From 1868, with the arrival of French immigrants and the first of the English settlers, the population of the colony steadily expanded. But with the historical records being, at best, so fragmentary, it is impossible to know precisely how many immigrants arrived. Several hundred, however, certainly passed through the colony, the vast majority leaving after just a few months (or even days), exhausted by the delays that they had experienced in Rio, Paranaguá or Curitiba, and frustrated by the lack of preparation made for them at what should have been the end of their long journey.[85] In 1874, when new British immigrants had ceased arriving, the colony's total population had reached 1130 people, of whom 391 were children under the age of ten. The resident population born in Britain numbered 260 (219 who identified themselves as "English" and 41 as "Irish"), not including their many Brazilian-born children. Immigrants born in Britain making up 23 percent of Assunguy's population were slightly more than the total of French and Germans combined (respectively 146 and 111 individuals, or 22.7 percent of the population).[86] By the following year, Assunguy's total population had risen to 1824, largely due to the arrival of more immigrants from Italy (202, whereas the year before there were just a few) France (a total of 338), from elsewhere in Brazil and babies born to earlier immigrants. With 262 people (221 "English" and 41 "Irish"), the colony's British-born had stabilized, but now made up only fourteen per cent of the total population.[87]

The British contingent was quite diverse in terms of social background, rather more so than that in Cananéia. Thomas Bigg-Wither was of the opinion that two-thirds of Assunguy's British immigrants were "town roughs and rabble" who even in the best of conditions stood not a chance of succeeding. Bigg-Wither based this estimation of his fellow countrymen on "their speech and their appearance", feeling that the Brazilian government had been deluded into somehow imagining that it was importing "English agriculturalists, well acquainted with the improved modes of farming as practised in their own land".[88] The reality, though, was somewhat more complicated. It is certainly true that few of the Irish – some of whom had been relocated from the failed Colônia Príncipe Dom Pedro, while others had been independently sent out under the auspices of the Brazilian consulate in Liverpool – had any obvious farming experience. As for the English, the majority were labourers (some with agricultural experience, and some without) who had been recruited in the small towns and villages of Dorset, Gloucestershire, Warwickshire and neighbouring counties; but others were immigrants recruited in 1868 in London and elsewhere in England, of middle class origins who were, in the words of one of their number, "indebted to the glorious descriptions of Meadows and Christopher", under the illusion that they were bound for an "El Dorado".[89] These immigrants were vital influences within Assunguy's British community, in part because of the deference that English labourers assumed before their social 'superiors', but also because these individuals were amongst the first to settle in the colony and had acquired valuable experience, both of pioneering life and of negotiating with officials locally and in Curitiba. Being better educated than most of the later arrivals and displaying a self-confidence that came with their social position, these immigrants took the lead in uniting their compatriots, organizing meetings, drafting petitions and making representations to the British legation in Rio de Janeiro on behalf of the community generally or to support individuals in particular need.

One of the most influential and most articulate of the English immigrants was Frederick Tertius Tigar who arrived in Brazil in 1868 at the age of thirty. Emphatically describing himself as "not one of the working class of England", Tigar was from a wealthy Yorkshire background, the son of a former mayor of Beverley, although by the time he reached adulthood the family had somehow lost their once considerable business interests, leaving the young man to fend for himself. After a period in the navy, Tigar decided to emigrate and was clearly attracted to Brazil by the fact that capital was not a requirement to acquire a farm. Tigar made considerable efforts to overcome Assunguy's many obstacles, displaying a determination to succeed that was matched by very few of the other English immigrants. For years after arriving in Brazil, Tigar maintained that he was not one of "the discontented English of Parana", having created "a rough but comfortable home", and he most certainly disassociated himself from those immigrants he referred to as the "idle vagabond class" whom he felt were giving the British an unfair reputation. Although Tigar was clear that he regarded his own position as rather different to that of most other British immigrants, he represented his compatriots as an elected member of the colony's advisory council, a dubious honour, he felt, complaining that he received no pay, although his duties involved travelling to Curitiba at his own expense, in one instance having to spend two weeks away during the planting season. Despite his frustrations, Tigar was sympathetic to the plight of the colony's directorate, regarding them as "more sinned against than sinning", feeling that they had no autonomy to improve conditions locally, acting solely as they did under the direction of superiors in Curitiba.[90]

Arriving in Assunguy at about the same time as Tigar were Charles and Caroline Tamplin, who left their home in London for Brazil following "a sudden reverse of fortune" and having for financial reasons ruled out British colonies. Already in their late forties and with five children when they settled in Assunguy, for the middle class Tamplin family it was a struggle to adapt to frontier life as pioneer farmers. The family shared the experience of lengthy delays before they were finally assigned an allotment, missing the crucial planting season. As a result the family's first crops were only harvested fifteen months after they arrived in Brazil. They survived in the meantime on the ten days' provisions that the early arrivals had to make do with, selling such personal effects as they had brought with them from London, living for months on end on a diet of boiled corn and salt and altogether "suffering the greatest privations and hardships".[91] Having sold the valuables that they had brought from London and, whatever the precise reasons, having nothing to return to, the Tamplins struggled to make the most of their Brazilian home, suffering tragedies similar to so many others:

These many misfortunes reduced us so much that we became heavily in debt. My Dear husband who was a strong man, fell into such ill health that it quite incapacitated him from work and these disappointments weighed so heavily on his spirits and on a frame already weakened by disease, that in little more than eighteen months from his appointment in the colony he sank under great misery. The opinion pronounced by the medical man being that he died from heart disease, greatly hastened by anxiety and privations.[92]

A persistent cause for complaint of the early colonists was that there was no Protestant clergyman located nearer to Assunguy than Santa Catarina, despite the many English and German adherents arriving in Paraná. A practical problem facing these immigrants was that marriages could therefore not be celebrated between Protestants – only Roman Catholic ceremonies were possible, as under Brazilian law civil weddings were not permitted and mixed marriages were only performed by a Catholic priest under the explicit understanding that the children of the marriage would be brought up as Catholics.[93] Only in 1875 was a Protestant church built in Assunguy, although it was little used due to a location far from most of the settlers and, in any case, a clergyman was not appointed until May 1877. Joseph Redman, who arrived in Brazil as

The Turvo Valley schoolhouse. (Tigar family)

an independent missionary, was a government employee, the Protestant parishioners being too poor to support him. Despite his official status, Redman usually went unpaid and, without any land to fall back on, he was even worse off than the *colonos* he served.[94]

For years there was also neither an English school nor an English teacher in Assunguy, something that was especially a source of anger amongst the 1872 and 1873 arrivals, most of who arrived in the colony as large family units. This clearly violated the terms of the contracts under which immigrants had been recruited. Immigrants were unable to register at the Brazilian school, leading Frederick Tigar to comment, "[a]s matters stand at present the children are running about wild, and in many cases, alas, the parents cannot teach their children being untaught themselves", people who were "in some cases almost as deplorably ignorant as savages".[95] When a school was eventually created for the English children, attendance was reported to be poor, as it served a community that was widely scattered and could no longer consider schooling as a priority – if indeed many members had in England.[96] An accomplished linguist, artist and pianist – amongst the precious items brought with the family from London was, according to family lore, an harmonium, somehow carried by mule along the Assunguy road – the widow Caroline Tamplin was considered the most cultured of the British immigrants and the obvious choice as a school teacher. She opened her school on 1 February 1875, attended by a few children who lived nearby. As with other government work, Caroline was in a constant state of dispute regarding payments, her monthly salary of fifty milreis rarely actually given and eventually reduced to a nominal fifteen milreis. Like many of the other immigrants who gave Assunguy a chance, she regarded herself as a prisoner of accumulated debts.[97]

Health

By and large, Assunguy was regarded as reasonably healthy, with no more medical concerns than would be expected of an isolated community. In early 1873, following an outbreak of illness amongst English newcomers to the colony, Charles Tamplin was appointed Assunguy's doctor. How this appointment came about is unknown: Charles was not medically qualified but either managed to convince the colony's administration that he had in fact been a practitioner in London – something that has been recorded in family lore – or the need for a doctor was simply too great to be concerned with qualifications. This appointment, however, helped the family financially, although they complained that Charles received just sixty milreis per month instead of the promised 150 milreis, and he held the position for ten weeks in 1873 until a doctor arrived from Rio.[98] From July 1873 there was also an English nurse, Joseph Renaudin, who distributed medicines, helped with operations – and claimed not to be paid the agreed monthly salary of 35 milreis.[99] Considering the number of immigrants arriving in Assunguy, it certainly appears there were few calls on the medical services. There were many other complaints regarding Assunguy, but the only medical issues recorded that affected British immigrants between 1868 and 1875 were the deaths of ten children during the especially hot summer of 1873 to 1874, said to have been the result of being permitted to eat unripe oranges and pineapples, a woman (Mrs Randall) who died of a haemorrhage of the womb and two men (probably including Charles Tamplin and Ubald Tetu, a doctor from Québec who had moved to Brazil with his family, apparently for health reasons) who died of tuberculosis believed to have been contracted before arriving in Brazil.[100]

*   *   *

Even given the most suitable of immigrants, neither Colônia Cananéia nor Colônia Assunguy could be said to have offered at all promising conditions for the development of farming communities. After the departure of so many of the newly arrived immigrants, the colonies continued to decline steadily, despite the fact that out of either desperation or determination many of those who remained were the more determined.

Chapter 8
The Collapse

MANY OF THE British immigrants due to be dispatched to a Brazilian state colony did not go beyond Rio de Janeiro or one of the Ministry of Agriculture's hastily organized holding stations. Many of these immigrants heeded the horror stories that they heard of the colonies from people they met trying to return to England or to get to third countries. Of those who actually made it to Cananéia or Assunguy, a large proportion made little effort to settle, arriving in a state of exhaustion at under-resourced and disorganized colonies that were totally unprepared for their reception. Typically these newcomers would remain in temporary lodgings for a few days or weeks, fleeing at the first possible opportunity. Of those who received land, many abandoned their forest clearings after a single season, incapable of investing energy into the allotments, unwilling or unable to accumulate more debt when they realized that they had no realistic prospects of establishing themselves.

In an environment far removed from that which they had known in England, the immigrants were quite bewildered by their situation. "The people consider themselves transported without deserving the punishment", reported a visitor to Assunguy, who recalled one desperate settler having said to him, "'oh sir, I would willingly return to England perfectly naked to get out of this.'"[1] Realizing that appeals for assistance were likely to be futile – and quite possibly risky – if addressed to a colony's director or the province's government, immigrants instead placed their faith in the ability of British consular officers and legation diplomats in Rio de Janeiro to act on their behalf, hoping that they would intervene before the minister of foreign affairs or possibly the minister of agriculture.[2] Complaining that "we are destitute of money or tools and can scarcely procure the necessaries of life", a petition dated 1 March 1869 and signed by 21 men was apparently the first such appeal sent from Assunguy.³ A similar document was drafted at a meeting in Cananéia on 22 January 1870, the petitioners pleading to be sent to Canada or New Zealand, insisting that they were not seeking charity from the Brazilian government, but only that contractual obligations should be met "and not leave us here in a forest in a State of destitution after making us such fair promises".⁴ But with considerable power vested in directors of state colonies, immigrants were wary of being labelled as troublemakers. William Robinson, an Assunguy *colono* explained that the director's "liberal wine cellar" made it easy to convince people to turn against others, as many had against him, as they were liable "to swear anything you may put in their mouths for a belly full".⁵

Many of the immigrants struggled for years to create homes and farms in the most difficult of conditions. Some sincerely hoped against hope that conditions would improve, that the colonies would receive more central government support and that road or shipping links would be developed to enable their produce to be profitably sent to market. Others fell into a state of lethargy, unwilling to return to England as failures but having neither the energy nor the means to start again elsewhere. Carlos Cenci, the American who had formerly reported to the Brazilian Official Agency of Colonization, was commissioned by the British government to produce a report on conditions in Colônia Cananéia. He felt the immigrants were easily intimidated by authority – or merely drained by conditions and events – seeing no point in raising their voices in complaint. According to Cenci, Brazilian officials investigating state colonies would typically phrase their questioning of *colonos* along the lines of, "'Well, Mr A., you are very much pleased with this Colony, are you not? And have you no charges whatever to make against the Director or any other of the Colonial officers?'" To such a line of questioning a *colono* would realize the futility of complaining, responding with a simple 'yes' or 'no', before returning to his allotment.⁶

As early as February 1869 there had been reports of scores of immigrants arriving back in Rio on a daily basis, not only from the utterly disastrous Príncipe Dom Pedro, but also from Cananéia and Assunguy. Determined to leave Brazil as soon as possible, having suffered "a broken spirit and degraded position", one of these returnees expressed feelings shared by many others. "As long as I live", declared James Hurst on arriving back in Rio from Assunguy accompanied by his wife and two children, "I shall entertain the deepest regret that I ever saw Brazil or had anything to do with its Government, and untruthful agents for having induced me to leave my home in England under false pretences with no better prospect than to return to save my life in a penniless condition."[7]

Immigrants abandoning Assunguy and Cananéia sometimes, like the Hurst family, left as individuals. A few had sufficient financial resources to be able to leave Brazil without help. One such immigrant was James Randall, an agricultural labourer from Somerset who went out to Brazil in February 1872. After his wife died in the following year, he found himself neither able to work on his land nor to take on occasional work from Assunguy's directorate as he had to care for his young children alone. Randall was unusual in being somehow able to raise the equivalent of £8 from selling his crops and profiting from the improvements that he had made to his 75-acre allotment. He then made his way to Rio and on to England unassisted.[8] But with any modest savings that immigrants may have brought with them to Brazil rapidly depleted, most resorted to selling their last few remaining personal possessions (usually nothing more than old clothes and shoes) and then set off for the capital with less than two milreis for an entire family, trusting that they would be given food and shelter on the way. Regardless of the fact that groups of immigrants returning from the state colonies became a regular occurrence, their arrival in Rio continued to send the resident British business and diplomatic communities into a state that bordered on panic, with accusations flying as to whom were most to blame – the Brazilian authorities or the immigrants themselves – for the embarrassing sight of destitute Britons begging in the streets and arguments as to what measures should be taken to stop this.[9]

One such group, comprising 46 men, women and children, arrived in the Brazilian capital on 14 January 1874, claiming that their decision to leave Assunguy was only taken after seeing a notice, credited to George Buckley Mathew, the British minister in Rio, that had been posted on a storefront in the colony. The notice was almost immediately torn down (although not before it had sparked what one of the *colonos* who remained in Assunguy called a period of "England mania"). It supposedly announced that moored in Rio de Janeiro's harbour was a ship, scheduled to leave on 19 January, which would embark immigrants wishing to return to England. With the returnees finding no such vessel awaiting them, the group sought shelter at the Casa de Saúde but, as the hostel was full to capacity, admission was refused, a further four hundred immigrants having just arrived from England and Germany.

Refusing to take either moral or legal responsibility for the predicament of these British returning to Rio, the colonization agency felt that the Brazilian government was showing generosity in offering to send them back to Assunguy or to another Brazilian state colony, aid that the group firmly rejected. British residents of the city intervened with offers of employment, but these also were turned down. Concerned that yet more returnees were on their way to Rio and worried that 55 British immigrants, newly arrived in Brazil, were about to be turned out of the Casa de Saúde having received reports of the conditions experienced by earlier British immigrants, George Lennon Hunt, the British consul, issued a statement urging immigrants to remain in the state colonies and denying the authenticity of the notice credited to Mathew.[10] Seeking to dampen expectations that any disgruntled British immigrant arriving in Rio would be automatically repatriated to England, George Preston, the British consular chaplain in the Brazilian capital, sent a letter to Assunguy's British residents:

TO THE ENGLISH COLONISTS AT ASSUNGUY

My Fellow Countrymen – I am sorry to have to write to you in what some may consider an unkind way, but I do so in order that you may not be deceived.

You must be aware that many families have left the colonies and made their way to Rio in the hopes of finding friends to pay their passage to England.

This they have done, but at great cost to the English residents of this city, and I now write to let you know that I think it impossible to raise another subscription, and therefore I advise you to remain in the colony and try to earn a living.

If you have any complaints to make, I shall be happy to receive them, and put them into the hands of those who will see that they are properly attended to.

If after this warning you still determine to come to Rio, I beg you to take notice that you must make up your minds to work in and about Rio, for which you cannot gain much above 2$ per day, for the

English residents cannot subscribe more money to send you home.
I am your faithful friend. – *George Henry Preston*. – British consular chaplain.

Rio de Janeiro – February – 4 – 1874.[11]

Meanwhile, William Scully – whose *Anglo-Brazilian Times* was forever at loggerheads with Hunt – attacked the consul's support for the British immigrants as being misdirected, irresponsible and sensationalist, and sent a local British medical doctor, William Fairburn, to examine the returnees. Dr Fairburn reported that he only diagnosed "trivial ailments, which time and acclimatization along with cleanliness would speedily cure". The children, he declared, were merely experiencing prickly heat, small boils and mosquito bites, for which the discomfort could be simply eased. This view was corroborated by the chief surgeon of a visiting British warship who expressed the opinion that any discomfort that the immigrants were experiencing was of their own doing, the result of "their filthy personal state, their economy in the use of soap and water, and the deplorably filthy condition in which they kept their rooms". As for the possibility of the immigrants starving to death, Scully insisted that this was patent nonsense, given the availability of work offered locally and of other assistance being provided by Brazilians and British alike. Perhaps unwilling to accept that conditions in one of the state colonies that he had been so recently keen to promote were so very trying, Scully saw the fault as lying with the immigrants, "a greater set of idle ruffians [who ever] infested any country", who were choosing to reject what he claimed was well-paid work, opting instead "to live as able-bodied paupers on charity", charges that Consul Hunt firmly rejected, pointing out that, apart from the British, there were 470 German, Austrian, Swedish, Polish and French immigrants recently returned from the state colonies, all having suffered similar experiences.[12] It was certainly true that it was not only the British who were leaving Assunguy: French immigrants were said to be departing for Argentina in particularly large numbers.[13]

For many months Scully was to remain extremely critical of both the departing British immigrants – whom he insisted had "fair prospects before them" – and British officials whom he accused of being "enemies of Brazil" wilfully stirring-up trouble, only explicable as the taking of offence that English emigrants were being diverted there:[14]

It is to the benefit of the English poor and English commerce that nuclei of British colonization should be successfully established in this country, in which life and property are safe, under constitutional monarchical institutions like those of England, from banditry and revolution. A few small prosperous nuclei will ere long induce spontaneous immigration in thousands to a country of settled and peaceful rule, whose delightful climate and many other natural advantages place it intrinsically one of the first in rank among the future homes of European immigration.[15]

While urging Britons still in Assunguy not to abandon the colony, Hunt's more immediate predicament was what to do with those destitute British subjects who had already reached the capital, in the height of what was a particularly scorching summer, and were now gathering in large numbers at the consulate "in a state of great misery", sleeping rough in the city's streets. At a meeting of prominent members of Rio's British community, Hunt appealed to his audience to take pity on their less fortunate compatriots, stressing the seriousness of the situation by pointing out that many were now sick, with children suffering "limbs in a diseased state from insect bites" and saying that urgent action was needed "to prevent these persons dying of starvation". It was not an option, argued Hunt, for the immigrants to remain in Rio, as the wages being offered them would not cover the city's high cost of living and, in any case, the tropical climate was "unfit for the outdoor employment of Europeans". In addition most of the immigrants rejected the very idea of becoming labourers in Rio, explaining that they had left England to establish themselves as independent farmers and were not at all interested in poorly paid and insecure work. Instead, Hunt succeeded in raising almost £600 from British residents of the city which, thanks to discounted passages offered by the Pacific Steamship Navigation Company, was sufficient to send home 96 married and sick British subjects, leaving healthy single men to work their passage to England or accept temporary work in Rio.[16]

Who should pay to enable British immigrants to return to England remained a source of friction within Rio's British business and diplomatic community and also between the British legation and the Brazilian government. As immigrants continued to arrive in Rio, it became clear that local charity could not be counted on as the basis to finance future repatriations. On a personal level, George Buckley Mathew, the British minister in Rio, wrote on 7 April 1874 that not only was "the state of our duped emigrants" absorbing most of his working time, but also that their misery had "ruptured my pockets".[17]

Whether diplomatically and politically shamed, or perhaps accepting some responsibility for the suffering that the British immigrants had endured in the state colonies, the Brazilian government was persuaded to at least to contribute towards passage money, allowing – even encouraging – people to travel to the United States, possibly in an attempt to head off the considerable damage that was by now being done in England to Brazil's reputation.[18] By late February 1874 – within a period of less than two months – thanks to the charity of British residents, the consulate sent home 170 men, women and children at a total cost of 8,552 milreis (approximately £900), and arranged for a further 252 immigrants to leave the country on passages paid for by the Brazilian government at a cost of 14,382 milreis (£1,500).[19] By April, British officials in Rio had dispatched to England five hundred immigrants, but seven hundred British men, women and children remained, with diplomatic squabbles continuing as to whom should take responsibility for them.[20]

George Buckley Mathew appealed to London for the Foreign Office in future to advance passage money for distressed British immigrants.[21] Lord Granville, the foreign secretary, agreed that if the Brazilian authorities refused to pay, the British legation could in future finance the repatriation of those immigrants considered in most urgent need, but only men with wives and families who were considered quite helpless. Single men and other able bodied adults were not to be offered assistance.[22] Although much of the financial burden on Rio's British community was lifted, a steady flow of Britons from São Paulo and Paraná arrived in Rio over the few next months and years. The repatriation procedure was now more cumbersome, first requiring the rejection of an application by the Brazilian government and then another lengthy wait for a decision to be made by the Foreign Office in London on the particular case.[23] British diplomats then had the responsibility to pursue the Brazilian government to reimburse expenses incurred for the relief of immigrants while in Rio and for their passage back to England.[24]

No matter how badly off the British immigrants may have been in Rio de Janeiro, at least they were under the watchful eye of people who had strong reasons for wanting something to be done for them. For British residents of the city there was the embarrassment of being associated with such a body of totally destitute fellow countrymen: it was feared that English-speaking men, women and children dressed in rags and reduced to begging could only harm the prestige of the established diplomatic and business community. For the Brazilian authorities, the arrival in the capital of so many people who not long before had been touted as ideal immigrants, was clear evidence of the view that state colonies were an expensive and poorly conceived and managed failure.

But for many who fled Cananéia or Assunguy, reaching Rio took many months. Immigrants trying to get back to Rio would generally first head for Paranaguá or Santos in the hope that somehow they would be granted a free passage on a northbound coastal steamer. Many of these people found themselves stranded for months in these small port towns, living – if, that is, they did survive – in the most appalling of conditions. News reached Rio of the hardships being endured by Britons stranded in Paranaguá: Levy Walsh exclaimed how desperate he was to return to "Dear Old England". Having endured eight months stranded in the town, he explained in a letter the straits to which he and his fellow immigrants had been reduced:

> I never saw Beggers in England in such Distress as the English Emigrants are in this Place. [....] We are Eaten up with all maner of Vermen Besides covered all with Soors. [....] May God Bless all the Emigrants that come in this Country. We are here half starved and Almost naked. [*sic*][25]

In response to such appeals, Charles Dundas, the British consul in Santos, was sent south to investigate conditions. "Rarely has it been my lot to witness such misery", reported Dundas, confirming many of the stories that diplomats in Rio had heard. "With scarcely any clothes to their backs, without shoes to their feet, or so much as a blanket to cover them, without food or the means of procuring food, if it was not actual starvation it was a very near approach to it", wrote the consul, shocked that British subjects were reduced to eating discarded scraps of raw *carne seca* and orange peel off the street. Housed in a barracks, some for several months, dozens of British were crowded together "without

8. The Collapse

heed to modesty". Offsetting possible accusations that the men amongst them were simply lazy, Dundas pointed out that even had there been work available, "many were incapable of availing themselves of it, owing to the state of their feet and legs from sores".[26] Criticism was unfair:

> Vagabonds and scum of the earth have been epithets applied to them. Of such there may have been and I regret to say that there were too many of that class who came to Brazil in the belief and hope that hard work was either unknown or unnecessary here. [But] certainly I saw among the few in Paranaguá as fine working men as ever I could pick from the best districts in England. If they were idle and worthless, as has been alleged, the treatment they met with at the beginning was not calculated to encourage them to be any better; rather would it demoralize them more.[27]

Employment opportunities in Paranaguá were very limited. The only possible unskilled work that was occasionally available to the Britons was as railway navvies, paying barely enough money to survive, even if they were lucky enough to be taken on, leaving labourers even less well off than they had been in England, a state that only very few were prepared to countenance.[28]

Ellen Godwin, was one of those immigrants stranded in Paranaguá visited by Dundas. Mrs Godwin's husband had died in Assunguy, leaving her alone on an isolated forest clearing to care for their baby. Unable to go out to work on her land while caring for the baby and reduced to a state close to starvation, Mrs Godwin sold what possessions she still had and left for Curitiba. There she appealed to the provincial president for support, but claimed that the authorities "turned me in the street with not one penny in my pocket and I begged my way to Paranaguá" where she was also forced to survive by begging.[29]

Frank White told a similar story. After his wife and one of their children died in March and April 1874, White left Assunguy for Curitiba with his five surviving children, from where they walked to Paranaguá. Arriving in a state of starvation, White fell ill and was taken to the town's hospital, while his children, left on their own, were "forced to beg from door to door" in order to survive.[30]

After being pressed by the British legation on the condition of the British immigrants in Paranaguá, the Brazilian Ministry of Agriculture agreed to pay for their passages home, though not before making it clear that this was an "exceptional humanitarian gesture".[31] For some this "gesture" came too late. Levy Walsh died of yellow fever soon after reaching Rio, whereupon the British minister, Drummond, approached the Brazilian foreign minister, the Visconde Caravellas, to ask that the rest of the family be returned to England as a matter of urgency. While waiting for a reply, Mrs Welsh died, whereupon Drummond upbraided Caravellas, remarking that had his ministry dealt with the matter promptly the children would not have become orphans. After their mother's death, the children were taken by English families in Rio and given work in an English-owned hotel where they were told that they could remain unless family in England chose to claim them.[32]

Concerned by the flow of complaints emanating from Assunguy and the steady stream of immigrants arriving back in Rio, in late 1876 a British consular official was sent to Paraná to investigate. While some aspects of his report were quite positive – that living conditions had improved slightly, that work was now being given at times of the year when farmers could more easily be absent from their land, that less public money was being wasted and that overall fewer complaints were being levied at the administration – the general tone was critical. It was noted that the road linking Assunguy with Curitiba was still far from complete and, with only part passable by pack mules, some of it had actually deteriorated in recent years. As for employment, it was pointed out that the British *colonos* felt paid work was being unfairly distributed and that supplies could only be purchased from stores linked with government officials. Above all, the immigrants who chose to remain in Assunguy felt that their commitment to the colony was not recognized. They were still expected to repay their accumulated debts for travel and other expenses, while those who were leaving effectively had their debts cancelled.[33]

The Brazilian minister of agriculture rejected most of these charges. He insisted that work was always divided fairly and regardless of nationality. Of the total of 581 local men contracted to work, the largest proportion (184) was French or French-Swiss, but the second highest figure was British (138). Not specified was how much work was actually given, nor how much work was also contracted to Germans and others from outside the colony. Rather unconvincingly, given the lack of progress over the course of well over a decade, the minister maintained

8. The Collapse

that there remained a firm intention to construct a "carriage road" to Assunguy "as soon as the financial state of the country permits". The minister accepted the word of Assunguy's director that no one in the colony was obliged to purchase from any particular storekeeper, which may have, strictly speaking, been correct, although in practice only those with government connections would have been prepared to accept the credit notes given to *colonos* for work done. As for cancelling debts accumulated for passage to Brazil and temporary housing, implements, seeds, and land in the colony, it was argued that such a request was entirely unreasonable and unfair to other immigrants. "There is no instance of a single colonist having been constrained to sell what he possesses to indemnify the nation and the Treasury", the minister assured Britain's chief representative in Rio. "On the contrary," he continued, "terms of payment [are] prolonged for all those who cannot settle their debt till such time as it may suit them." Generally, then, complaints were dismissed by the minister of agriculture as "absolutely wanting in foundation [...] the Ministry under my charge zealously applies itself to the prosperity of the colonies of the state".[34]

The British immigrants in Assunguy considered their situation to be increasingly dire, regardless of how things may have seemed from Rio. Since being appointed Assunguy's Protestant pastor in 1877, Joseph Redman had worked hard to assist people to leave the colony, even requesting that all the British *colonos* be evacuated.[35] Redman believed that many of the immigrants were not as "steady, careful, industrious or moral as they ought to be", and therefore chiefly had themselves to blame for their lack of success in Brazil. But there were also many whom Redman considered as having "striven nobly with difficulties which they find insuperable", people whom the pastor felt especially deserving of sympathy and support.[36] Thanks to Redman's intervention, the British Legation agreed, for example, that Gloucestershire emigrant Emma Pullen – whose husband had died in Assunguy – and her eight children should be helped to return to England, although a request that four mules be hired to carry the family's belongings to the coast was firmly rejected. Another appeal was made on behalf of John Sellwood, a widower who was too ill to work his land, and his two daughters – despite one of the young women being said to be of "bad repute".[37] The desperate situation of recently widowed Hannah Randall and her seven children was also brought to the attention of Britain's diplomatic representatives. Mr Randall had fortunately been "a very steady, trustworthy man" perhaps making it easier for the Legation to agree to send Mrs Randall and her five youngest children home to England, but her two adult sons were told that they would have to work their passages.[38] The Millers were another "worthy and honest" family who deserved – and were given – assistance to leave the country, Mr Miller – a former soldier – having served his country well in both the Crimea and the Indian Mutiny.[39]

Despite the exodus, Frederick Tigar – one of the first immigrants to take up land in Assunguy and the unofficial representative of the British *colonos* – had managed to remain cautiously optimistic that the colony could be made a success. But by 1875 he was fast giving up all remaining hope. "As the case stands," he wrote despairingly to the British minister in Rio, "we are simply 'white slaves' – we work our lands for Brazilian task masters and leave our children for their heritage ignorance." He was particularly harsh in his criticism of provincial officials in Curitiba, referring to them as "the bane of the colony" who somehow "wish[ed] to place people into serfdom". Pointing out that "debts hang over us like the sword of Damocles", Tigar urged that outstanding amounts should be cancelled "as an atonement for the wrongs done". Despite Assunguy's generally miserable plight, Tigar felt that it was not altogether too late to raise the position of the *colonos* "from serfs to freedmen". But his frustration was clear. "If you require anything of a Director, Engineer, or a Person in Office, it must be cap in hand", wrote Tigar, adding, "we are truly the servile offspring of the free."[40] Assunguy's conditions had become utterly "wretched", with all the immigrants who could do so now abandoning the colony, encouraged, so Tigar believed, by the Brazilian government. Its policies respecting the state colonies were "inimical to our gaining a livelihood, granting no work and suppressing schools". Tigar reported heartbreaking scenes almost identical to those that Thomas Hardy created for his fictional character, Angel Clare:

> I have seen here which I never expected to have seen, men crying, begging for work to support their families, and it has been refused. There was but one feature pervading the colony, misery, destitution, the same events happening each year.[41]

When *colonos* brought complaints to the colonial administration, warning that they might have no choice but to leave, according to Tigar the reply would merely be: "the road is open, there are plenty of Brazilians ready to take up land".[42] Other immigrants also directed their contempt towards Assunguy's administration. William Robinson – also a veteran of the colony – claimed that he still believed that the policies framed in Rio de Janeiro relating to state colonies were sound ones and that it was their execution on the ground that was at fault. Although desperate to leave Assunguy, Robinson insisted that he would be perfectly happy to be transferred elsewhere in Brazil, convinced that "there is not such a liberal Government in the whole Universe".[43]

Not until 1879 did Frederick Tigar make a final decision to leave Brazil. Seeing how construction work on the road to Curitiba had completely ground to a halt and with opportunities to sell his farm surplus worse than ever before, Tigar reached the assessment that "I have struggled hard to succeed, fighting against adverse circumstances, but am obliged to own up that I am beaten".[44] Used to sending appeals to Rio on behalf of other immigrants requiring help to leave Brazil, Tigar turned to Joseph Redman, Assunguy's English pastor, who wrote a testimonial in his support: "[h]e is of a good family in England, a most steady, respectable and industrious colonist". The Foreign Office agreed to the request but, as a consequence of sickness in the family, the Tigars – Frederick, his wife Kathleen and their three young children – remained in Brazil for over one more year, their living conditions steadily deteriorating. Despite twelve years' hard work, the Tigars' farm was of little value; the sale of land, house and personal effects only raising 70 milreis (£7 6s.), just enough to pay to cover the expenses to take the family by mule and cart to Curitiba and onwards to Paranaguá.[45] On 9 November 1880, the Tigar family sailed from Rio for England.[46]

The Tigar family's departure must have been especially painful for Frederick's wife, Kathleen, the daughter of Caroline Tamplin – Assunguy's temporary British schoolteacher, a rather cultured woman trapped in a Brazilian backwater. Having left London for financial reasons and lost her husband in the colony, for Mrs Tamplin at the age of sixty there was nothing to return to in England. Described by another British immigrant as a woman who "understands music and has had an education something more than ordinary and with a high flow of eloquence", she was able to ingratiate herself with the directorate of the colony, helped perhaps by her conversion to Catholicism.[47] But connections hardly resulted in her being obviously privileged: she was forced to walk to Curitiba to obtain payments owed to her as schoolmistress and had her pay drastically reduced, placing her in the same situation as countless other less well connected immigrants. In 1881, a few months after the Tigars left Brazil, Mrs Tamplin moved to Curitiba. There she devoted her life to surviving family members and to the Church, attending Mass virtually every day. Giving up any claim to land in Assunguy, Mrs Tamplin instead made a "precarious subsistence" by teaching music and languages, living from week to week at a rate of pay of two milreis (less than a shilling) an hour, with pupils difficult to find and having to defend herself from the jealousy of local teachers.[48]

The personal circumstances of the Protestant pastor, Joseph Redman, were no better – or even worse – than most of the people he ministered to and assisted. Redman had to survive for many months at a stretch without pay and Assunguy's administrators refused to give him land to cultivate, despite his contractual entitlement.[49] Frederick and Kathleen Tigar drafted a petition, signed by 103 other British settlers, asking the British Legation in future to support their pastor and to help Redman pursue a claim for the pay owed him since the colony's director had found an excuse to suspend his salary.[50] Redman had been accused of breaching his contract by only performing services on alternate Sundays. While Redman accepted that this was indeed true, he explained that he had to travel considerable distances from the colony's centre to visit outlying homesteads. Appealing for help, Redman travelled all the way to Rio de Janeiro where British diplomats managed to persuade the Ministry of Agriculture to reinstate his salary.[51] But when Redman returned to Assunguy, the colony's officials still refused to grant him back payments, and he found himself virtually trapped. Without even any small interest tied up in land, there was no possibility of clearing the debts that had accumulated since had been salary was suspended. The more pressing problem was that "the little ones are in such a desperate state for want of money". Redman and his family were entirely dependent on handouts of food from other immigrants.[52]

Although torn by not wanting to abandon the other British immigrants still in Assunguy, Redman became increasingly focussed on the need to leave Brazil – "we are enduring the greatest hardships and privations in the family", he declared, appealing to the British consulate to help with passages back to England and with his claim for a year's back pay.[53] Redman's case was referred to London where the Foreign Office responded that the Ministry of Agriculture should be required to take responsibility for the family's repatriation, while accepting that if the Brazilian authorities refused to help then the British Legation could assume responsibility.[54] Redman received reassurances that the Ministry of Agriculture would in fact give him his back pay, something confirmed repeatedly to the British Legation by the Ministry of Foreign Affairs. The wrangling, as to whether the debt had been paid in full or not, continued for years, Redman claiming some 30,000 milreis (£4,500) for lost earnings and damages from the Brazilian government.[55] Even the Catholic priest in Assunguy supported Redman, lending him money to go to Curitiba to pursue his claim.[56] Meanwhile, Redman's appeals to Rio took on a more desperate tone: "my family is threatened with slow starvation", he wrote in one letter, claiming that they were under constant threat of disease and death due to the poor food that they had to consume.[57] His health failing, Redman reached the conclusion that that he had no option but to leave the colony and "remove my family to some civilized part of the globe".[58] In 1883 British diplomats in Rio agreed to pay for the Redmans to travel from Santos to Southampton but the family remained in Assunguy, as Joseph Redman was too ill to walk and the children were getting weaker by the day.[59] On 2 August 1883 one of Redman's sons died "of starvation". He reported that two others were "fast failing", being so emaciated that he feared that, even if they recovered, they would be invalids for the rest of their lives.[60] On 1 February 1884 the family arrived in Santos, a scene described by the British consul there:

> They were all in the most distressing condition of destitution, without shoes or stockings, and with only a few rags to their backs. Mr Redman's condition was really most deplorable, and I was in fear of his life from the moment of his arrival. The whole family seems to have been suffering from a long period of want of food.[61]

The consul supplied the family with some inexpensive clothing and placed them in second-class cabins ("I do not think the third class at all a fit place for an educated Englishman to be put, as only rough dirty Portuguese go there", the Royal Mail's agent had advised) on the steamer La Plata. It was hoped that the voyage to England would revive them.[62] Arriving in Northamptonshire at the beginning of March 1884, Redman continued obsessively to pursue his claims against the Brazilian authorities.[63]

The population of Colônia Cananéia was even less stable than that of Assunguy, with many of the immigrants not leaving the colony's port. Of those who persevered, most were from the 1868 and 1869 contingents, these immigrants having at least benefited from an efficient issuing of land. For the British immigrants who did not flee at the first available opportunity, the lack of someone such as Assunguy's Frederick Tigar or Joseph Redman to take the role of community leader and representative, made appeals for help both difficult and likely to be ignored. Most of the British immigrants who decided to leave the colony made their way to Rio or Santos as best they could, there to appeal directly to the British consuls for help in leaving the country.[64] The population soon dwindled to just a few British families, who survived thanks to occasional handouts from the pockets of British consular staff or of family members (themselves struggling in Santos or elsewhere) and from the fact that their land was of good quality, although still without a market for any surpluses.[65]

Of the English and Irish who arrived in Cananéia, it was the Davies family, recruited in Birmingham, who remained the longest: despite being barely literate, James and Bridget maintained contact with British diplomats in Rio for years. The explanation that James gave for having remained in Cananéia when most of the other settlers had long-since given up and left was simple: "my family increased in the colony so much that I could never rais a nough mony to lave the colony i have six sons and two daughters sir" [*sic*].[66] Two sons from the family went to Santos where they managed to save some money but, not possessing a skill, decided to return to farm in Colônia Cananéia. The young men wanted to purchase some land with their savings and James appealed for help with legal matters "as we cannot write Brazilian".[67]

* * *

From the outset it was clear that neither Cananéia nor Assunguy offered promising conditions for the development of immigrant-based farming communities. But given the colonies' many problems – such as their topography, poor external and internal communications, lack of markets for any agricultural produce and under-resourced administrations – it is a wonder that any of the immigrants persisted for any length of time. While most of the immigrants did, in fact, leave very soon after arriving in Cananéia and Assunguy, what is perhaps more surprising is that some remained. In some instances they stayed for many years or for the remainder of their lives, perhaps due to the strength of individual determination or to a lack of any apparent alternatives combined with exhaustion.

Part V

Conclusion

ON THE FACE of it, encouraging Irish and English to emigrate (and in the case of many to re-emigrate) to Brazil made perfect sense to all those involved. With issues of labour, slavery and race so intertwined in the minds of its elite, Brazil stood to gain from the addition of an apparently vast reserve of white northern Europeans desperate for new opportunities. Additionally, it was assumed that English-speakers – people closely associated with the perceived progressive influences of Britain and the United States – would be hard working and would bring with them skills and introduce modern farming methods to publicly-owned land crying out for cultivation. Catholic Irish immigrants would, it was felt, integrate into Brazilian society especially easily, but Protestant immigrants, whether Irish or English, were also welcomed – indeed some Brazilians considered them to be even more industrious and therefore more desirable.

For promoters abroad, Brazil was a land of boundless opportunities, whether offering a venue for a Utopian Irish renaissance or simply providing the prospect of a more secure future in a country where immigrants would not suffer religious or ethnic persecution, would be free from constant fears of unemployment and eviction from their homes and would have their own land to cultivate. Compared to other available possibilities, the descriptions and terms of settlement must have made Brazil seem an earthly paradise.

But the reality, as almost all concerned would quickly discover, was quite different. Colônia Príncipe Dom Pedro and, although perhaps to a lesser extent, *colônias* Cananéia and Assunguy, from their very inception lacked any realistic chance of developing into self-sustaining, let alone dynamic, communities. The steep, wooded, hillsides and narrow valley floors that made up so much of their territories meant that agriculture was bound to be extremely difficult, even hazardous. Local infrastructures were hopelessly inadequate and markets distant, difficult and expensive to reach, not cost effective when attempting to sell the low-value agricultural surplus of a Brazilian peasant farmer. The local administrations of the colonies were at best let down by central government in far-off Rio de Janeiro, while at worst they were inept or even utterly corrupt, with the irregular flow of funds earmarked to support the struggling *colonos* rarely reaching the intended beneficiaries.

Controversy surrounded the suitability of the immigrants, only some of whom had agricultural experience. Rather than accept any culpability, the Brazilian authorities and many supporters of recruitment found it convenient to place the blame for failure entirely on the immigrants, accusing them of refusing to do anything to help improve their situation. The frequently voiced claims that "cripples, invalids and persons wholly unfit for agricultural pursuits" were sent out from England and that the New York recruits were nothing more than a collection of drunkards, "vagabonds and ragamuffins" had some bases of truth but the entire body of arrivals could not fairly be described in such terms.[1] Apart from blaming the immigrants themselves, the Brazilian government also lambasted their English agents who were accused of having made false claims in support of prospects in Brazil, unduly raising expectations.[2]

But the truth regarding the background and character of the immigrants was much more complex. While agriculturists amongst them may well have been a minority, a lack of farming experience would not necessarily have been a significant problem, had they been installed in less isolated and better funded and supported colonies. In the view of William Scully, the abilities required of pioneers dropped into a dense forest – where, without guidance, they would have to construct a shelter, clear and plant the land – were "to be strong and handy, industrious and persevering, patient and inexhaustibly hopeful, healthy and capable of stomaching beans, mandioca, farinha and dried beef of more than doubtful quality".[3] This was certainly true, but another requirement of immigrants in isolated agricultural colonies was to be able to accept and adapt to living as self-sufficient, peasant farmers, capable of enduring the rigours of an existence as pioneer settlers. While the magnet of cheap land was an important lure, most of the immigrants were quickly disillusioned. They discovered that, although living costs were inexpensive in comparison with England or the United States, without even basic infrastructure or external links most hopes or plans were unrealistic. Many of the English immigrants were tempted to Brazil by notions of economic independence based on land ownership, yet at the same time being assured that they would have the security of employment on public works. Most of the immigrants, however, quickly discovered that the apparent attractiveness of the Brazilian package was an illusion.

Apart from the writers of the enthusiastic letters reproduced in English newspapers during the time that agents were most actively recruiting for Brazil, British immigrants were almost always critical of the conditions that they found. Many of the surviving records take the form of letters and sworn depositions directed to British diplomatic and consular representatives. Of course, satisfied immigrants would have been far less likely to contact the British consulate and, as such, it could be argued that the testimonials that survive should be considered only part of the truth. That said, the consular officials investigating immigrants' conditions were always at pains to interview as many British subjects as possible, and never found more than a few satisfied *colonos*. Whatever the motivations of apparently contented letter-writers – genuine enthusiasm born of a brief time when conditions were encouraging, desire to encourage familiar faces to join them and build larger English-speaking communities, or pressure from Brazilian officials – from early 1873 hopeful letters were few in number, the previous authors having either left the colonies or changed their opinions.

Following the 1868 recruitment efforts, Brazil did not re-emerge as a destination for British emigrants until 1872. In July of that year, *The Labourers' Union Chronicle* endorsed Brazil as a suitable destination, declaring that the country's great attraction was the "certainty of securing an independent and prosperous position as a freehold cultivator, while in the meantime he is able to gain a satisfactory livelihood".[4] But within months, reports in local and national newspapers adopted far more critical tones with, for example, *The Wilts and Gloucestershire Standard* reproducing in February 1873 the following commentary on the unsuitability of Brazil as a destination for English emigrants:

> How different is this from their being sent to some country where their native English tongue salutes them on landing, and surrounds them with a familiar, cordial charm of their new abodes; where they

find their own ideas, their own habits and religion; where, if they are not well off in one place, they can move to another, unstopped by the insurmountable impediments of a strange tongue, strange life and customs, and a jealous people, such as in the transatlantic Portugal must hem them in. Can anyone, for a moment, believe that such a preposterous and unnatural anomaly – an isolated body of poor English labourers, surrounded by a race utterly foreign to them – is one that can prosper, or ought to be attempted?[5]

Reports of the hardships and sufferings being experienced by English emigrants in Brazil were by now circulating widely at home, with returnees speaking at public gatherings and letters being reproduced in newspapers from correspondents whose hopes and expectations had been brutally dashed. Typical of these letters was one from George Arnold, an agricultural labourer from Oxfordshire, who offered an entirely negative slant to the exoticism of Assunguy:

Brizilens don't reacon to eat Bread and the Language is Portugue [...] the hills are all over trees and rubbish there is no Grass at all that is good for anything and there is bits of Indian corne in some odd places. But up in the Collene there is no Chance of getting their Crops for Whild pigs and the Monkes will fetch it all in the nights [...] there are plenty of Orange trees and figs and Benaners in some places that they took us but there is none here and Coffee takes 5 years to get a crop it grows on trees and there is plenty Slaves hear snakes and Grass Hoppers and toads nearly as big as Hedge hogs you don't want no more ground than you can dig ploughs is no use up Mountains.[6] [*sic*]

But even as negative reports were filtering back home, Warwickshire agent Thomas Alsop claimed he was still receiving letters from emigrants "who advise their friends to come to the land of Canaan". Alsop continued to maintain that Brazil was "the *best place* to go to", adding that it defied logic for the Brazilian government to be investing considerable sums of money to assist English farm labourers to settle in agricultural colonies if there were no reasonable prospects for them:

It is absurd to say one or two [emigrants] have failed, or that fevers are known at Rio, or that there are snakes. Who would recommend a foreigner to settle in England if he was to be guided by our fogs and our pauperism?

The "tide" of emigration to Brazil, insisted Alsop, would not be stemmed by the "grumbling" of just a few Oxfordshire men whom he accused of being the source of what he insisted was simply malicious gossip.[7]

As Alsop continued his recruitment drive, critics of Brazilian emigration were raising their voices more forcefully. An anonymous writer in *The Royal Leamington Spa Courier and Warwickshire Standard* warned Warwickshire farm labourers that they would be giving up all that was familiar – "their cottages, their beer and food" merely to support Alsop's personal "speculation" efforts. Immigrants in Brazil were "used worse than beasts", and anyone going there could expect to have to sleep on the ground, exist on a diet of nothing more than black beans and rice, have but filthy water "to cool their parched throats", suffer "burning fever" and endure a climate that would "peel the skin from their bodies".[8]

Feelings towards Alsop grew angrier. In Napton, the agent's home village, the parents of Thomas Sheasby received a letter from Cananéia in which he wrote, "I think Mr Alsop served me very nasty", and expressed upset at being abandoned by Alsop when he left the colony. Sheasby explained that he was trying to return to England, that not a single immigrant would remain in the colony if they could possibly leave and that already fifteen British had died as a result of a lack of food or smallpox.[9] Expressing concern that Alsop was planning to escort another group to Brazil, Sheasby urged his parents to do their utmost to dissuade anyone who was being tempted by the offer.[10]

On 3 January 1873 a large meeting was held in Napton's Crown Inn to discuss the Brazilian situation. In heated exchanges, the gathering heard of immigrants "escaping" from Cananéia and that clothes were being sold to purchase rice. Alsop dismissed these charges, insisting that food was plentiful and all men, women and children in the colony received generous subsistence allowances.[11] While the Napton gathering was well ordered, in other places where feelings were running similarly high, control was difficult to maintain; a particularly "riotous meeting" was said to have taken place the following month in Banbury, Oxfordshire, where many emigrants for Brazil had been recruited.[12]

Within a matter of months, there remained no possibility of recruiting in Warwickshire any more emigrants for Brazil, and soon after it also became impossible to attract people in the West Country. In September 1873 the vicar of Napton wrote in *The Times* of London that

150 emigrants had just returned from Brazil. "They can hardly realize", he said, "the joyful fact that they have reached their homes at last, and are strongly in their resolution never to be tempted out of old England again."[13]

It is quite likely that Joseph Arch, the leader of the National Agricultural Labourers' Union, also regretted having ever heard of Brazil, later denying any involvement with "the Brazilian mess". But Arch recalled how, at meetings that he attended to promote Canada as a destination, he often faced critics of union involvement with emigration who would call out, " 'Oh, be careful, be very careful about sending men to Canada. Look at the Brazilian fiasco; take warning by that.' "[14] While it may be strictly accurate that Arch did not personally encourage emigration to Brazil, there can be no doubt that there were definite links with the NALU. Apart from the promotional articles that appeared in the NALU-affiliated newspaper, *The Labourers' Union Chronicle*, it would certainly have appeared that Brazil was being endorsed when the country's agents were invited to speak at union meetings. And although the Gloucestershire union secretary Frederick Yeats may have recruited for Brazil in a private capacity, it must have been difficult for people to separate this from his official position, not least because he was signing men up as NALU members in order to enable them to declare themselves agriculturalists when applying for a passage.[15]

Despite Arch's protestations of innocence, critics of the NALU were able to make use of the disaster. One correspondent to *The Royal Leamington Spa Courier* referred to "the deluded labourers [who were] slow to perceive that by listening to the wiles of Messrs. Arch, Taylor, Alsop, and Co. they are exchanging a comparatively comfortable position for one of thraldom, misery, and want". The writer ridiculed these men's experience – placing the name of Brazilian emigration agent Thomas Alsop alongside those of the union leaders Joseph Arch and Henry Taylor – and enquiring, "who made them judges of the most desirable place for emigration?"[16]

Henry Tayor, a Leamington carpenter and NALU general secretary, also denied any official responsibility. He insisted that "neither the union, nor its agents, were in any way responsible for the statements made to those who were persuaded to go to that country" and claimed that he personally had always believed and argued that Brazil was not a fit place for English labourers. Nevertheless, he remained reluctant to condemn Brazil outright, saying in September 1873 that some families were doing well, far better than they could ever have managed in England – something that he believed boded well for future plans to assist labourers to emigrate to countries that in his opinion held out more promise.[17]

Regardless of its actual links with Brazilian emigration, it is certainly true that the NALU was sensitive to the issue. For example, at a meeting in the Warwickshire village of Hampton Chapel in the evening of 18 September, 1873, "that much-to-be-regretted scheme of emigration to the Brazils" was mentioned, although speakers were at pains to point out that the scheme "was not supported by the Union, which left men to form their own opinions, and act as they liked". But the meeting went on to consider prospects in Louisiana, the guest speaker offering homes to ten thousand "honest, industrious, English working people". There, the newspaper assured its readers, Warwickshire agricultural labourers would find a home where they "need not suffer under the iron heel of the oppressor, where the necessaries of life could be obtained by all willing to work, where the workhouse was not always staring one in the face, and where a man knew that labour would result in prosperity". With officers of the union presiding over such meetings, it must have appeared that at least some official weight was being given to what otherwise might have appeared highly speculative ventures.[18]

Beyond union circles, puzzlement was initially the dominant reaction in England to the Brazilian emigration episode, with *The Times* in April 1873 stating "the whole affair is so extraordinary as to be in some respects inexplicable".[19] In reaching for an explanation as to why a prospective emigrant would opt for Brazil when there would seem to be so many more plausible destinations, the editorial suggested that it could only be due to "the singularly passive and credulous character of the English agricultural labourer" and that "it is their helpless and gregarious nature which exposes them to careless usage, and even invites it". Readers of this Establishment newspaper were told that emigrants had been tempted by offers above their station: "regular employment, such as not to tax their strength and health, good treatment, and a high rate of wages […] the combination of profit, comfort and security which is seldom attainable". Sensing rural discontent at home, union "agitators" gathered "like vultures scenting a prey" and at the same time Brazilian interests had appeared on the scene, taking advantage of what *The Times* portrayed as simple, gullible folk.[20]

The English surveyor Thomas Bigg-Wither, who travelled extensively in Paraná in the early 1870s, talked with many of the immigrants. "That the agents had lied to these poor people in England, I have not the slightest doubt, the testimony of the settlers being overwhelming on this point, but that the Brazilian Government itself in any way connived at the deceptions practised, I can scarcely think likely." Despite acknowledging that "official morality in Brazil is certainly not held in very high esteem", Bigg-Wither could not believe that a government would contemplate "so much money [being] thrown into the sea" in an action that would only serve to harm Brazil's reputation.[21]

While pouring scorn on the very notion of Brazil as a suitable destination for British and Irish emigrants, others felt that they could understand why at least they were sought. "The object of the Brazilian is perfectly plain and comprehensible," explained one fierce critic, "he wants work done, and has himself an innate personal aversion to doing it."[22] The pseudonymous writer asserted that it was not surprising that the immigrants had been so poorly treated in a country accustomed to the institution of slavery and where labourers were considered a mere commodity. Brazil, he argued, could never be a suitable place for British to settle, other than for those going out on short-term contracts. Nor, indeed, should it be necessary:

> Between Canada and the Cape, Vancouver's and the Falklands, New Zealand and the Himalayas, Belize and Pegu, one should have thought there was sufficient work cut out and elbow-room enough, in all conscience, for our surplus Islanders, without their flying laws, religion, rites, and mother-tongue, in the depth of the damp Sertão [...]
>
> [...] If the steaming woods of the torrid zone are to be the habitation of the ideal man of the future, that man must be the Nigger, or at least more allied to the ape than we are. But if there are men who still have dreams of sitting under their own palm-trees, smoking their own fresh-plucked tobacco, have we not tropical lands enough, without trying to graft blackthorn upon banana, the Saxon on Mulatto or Iberian?[23]

Alluding to the notion of contributing British immigrants as part of a broader mission of 'whitening' being advocated by elements of the Brazilian elite, this critic urged that the established British–Brazilian relationship should remain firmly one of business, not straying beyond the "Platonic", as it would otherwise result in "a progeny of Creole or even mongrel Britons, effete examples of parental folly".[24]

Official British attitudes alternated between sympathy toward the plight of the emigrants, embarrassment at the sight of destitute English and Irishmen on the streets of the Brazilian capital, frustration at the Brazilian government's constant shifting of responsibility for the whole affair and concern over the expenses incurred in assisting the many distressed British subjects.

George Buckley Mathew – the British minister in Rio during much of this period – was consistent in his condemnations of the entire episode. In the introduction to an 1873 British parliamentary report on the subject, Mathew wrote:

> [It] is beyond question, the persons by whose extravagant statements and promises these unfortunate persons were induced to emigrate, were the paid agents of the Brazilian Consular functionaries in Liverpool or elsewhere, I cannot feel but convinced that the Brazilian Government are both legally and morally responsible for their losses and sufferings, and that any expenditure that may be ultimately inevitable for their return to England should be borne by that Government.[25]

Charles Dundas, the British consul in Rio de Janeiro in 1874, felt that the accusations often levied that the immigrants were good for nothing and lazy were entirely unfair, insisting that it was absurd to suggest that the immigrants actually "prefer loafing or starving [to working] as though that were quite a pleasant sort of life in Brazil". He too pointed the finger of blame at the agents in England and Brazilian officials, most of who were entirely unfit for their duties:

> Having been used to slaves all their lives they know nothing of handling British workmen. The idea seems to me to be to import into the country then upon the slightest provocation to break faith with them and turn them to graze like so many head of cattle to feed upon whatever pastures they can find.[26]

In contrast, Carlos Cenci, the American commissioned by the British government to undertake a series of reports concerning immigration, felt some sympathy for the predicament that Brazilian authorities found themselves:

> The Brazilian Government has imported a raw material which it is unable to utilize, of which it cannot rid itself, except at great expense and with much discredit, and the housing of which involves a constant outlay from which it can expect no return.[27]

In London, parliamentary questions were occasionally asked, and ministerial statements made, condemning Brazil as a place suitable for British emigrants.[28] Drawing on the experience of the more established British community in Argentina, a Foreign Office minister quoted the British chargé d'affaires in Buenos Aires to suggest that the outlook for Britons in Brazil was not likely to be encouraging:

> The English emigrant will find no encouragement; no similarity of language, habits, or religion; no liberal land laws, no economical and ready collocation on tracts of land ready and marked out, no ready access to markets for the sale of produce, and but scant and merely nominal protection for life and property.[29]

The Brazilian government adopted an extremely defensive posture. The minister of agriculture, Joaquim Antão Leão, blamed the problems of Colônia Príncipe Dom Pedro on poor quality immigrants carrying with them too high expectations and forbad "in the most absolute terms [the Brazilian government's] agents in Europe to make promises of exaggerated advantages and benefits, with the purpose of inducing immigrants to leave their country".[30] Even so, in 1876 – fifteen years after Assunguy received its first settlers and eight years after the British began to arrive – the Brazilian Ministry of Foreign Relations claimed that "the wants of colonists have for the greater part been attended to", stating that the construction of bridges and roads was well advanced, and that churches, schools and hospitals were almost finished.[31] To accusations that the state colonies received insufficient support from the government, the Ministry of Foreign Relations responded that it was "not indifferent to the welfare of their colonies, and [that the *colonos* are] attended to their comforts, as experience goes on teaching them".[32] While no doubt intending to sound reassuring, the statement implied that both the colonial authorities and the immigrants were learning through a process of trial and error. For pioneer settlers this would normally be regarded as a natural state of affairs, but without either expertise or finances to draw on, such a process was always likely to end in failure.

Some of the information being received by the central government in Rio de Janeiro from sources in the provinces supported this aggrieved stance. An anonymous correspondent from Assunguy writing in early 1870 in *O Dezenove de Dezembro* – a newspaper that served as the mouthpiece of the government of Paraná – confidently declared that hundreds, in future perhaps thousands, of European families would settle in the "fertile forests of Assunguy", attracted by the area's supposed capability of producing all temperate European crops including wheat, rye, barley, wine, flax and potatoes.[33] Even in 1874, when the exodus from Assunguy was at its height, the provincial government was still boasting of the colony's "extremely fertile land" upon which a hard working immigrant could achieve "incalculable profits". If there were problems, these were blamed on Brazilian agents in Europe and their lack of scruples in recruiting "lazy immigrants with bad habits".[34] In Assunguy itself, a Brazilian defender of the colony claimed in the same newspaper that, owing to its fertility, there was every reason to expect "a great future".[35]

Whether such opinions were held with any conviction is impossible to know. Certainly the weight of evidence conspired against them, and from the earliest years disappointment and criticism were also being voiced by official sources in Paraná. As early as 1866 the director of Assunguy acknowledged that without proper means of communicating with the outside world "it is impossible that [the *colonos*] will overcome the difficulties of poverty and disheartenment over which they struggle".[36] In 1870, two years after the arrival of the first wave of British immigrants, the *Dezenove de Dezembro* commented that Assunguy "resembles a granary with closed doors", but by the middle of the decade the newspaper felt "forced to confess" that the colony was failing due to lack of progress on the road.[37] By 1879, Assunguy's director mused that the colony could have developed into an important city – the business centre for the north of the province of Paraná – had a decent road to Curitiba been constructed.[38]

But even in official circles, such optimistic opinions were questioned and even seen as complete fantasies. The Ministry of Agriculture stated in 1872 that the construction of a tramway between Castro (a small town in the interior of Paraná) and the coast at Cananéia via Assunguy would be "indispensable" for the development of the entire region.[39] A detailed report, commissioned in 1873 by the Brazilian government's Official Agency of Colonization, went further, being devastatingly critical of all aspects of Assunguy, not only of communication difficulties but also the colony's location, terrain and administration.[40] Even Paraná's government was forced to accept reality. While pinning some blame for Assunguy's failure on successive administrations of the colony, even admitting to financial irregularities, the provincial government also accepted that local topographic conditions were unfavourable to European settlement.[41]

In line with the fluidity of his relationship with the Brazilian government, William Scully – who had been a key figure in the Sociedade Internacional da Imigração that had been established to promote Brazil as a destination for immigrants – was inconsistent in his views regarding state colonies. In the immediate wake of the collapse of Colônia Príncipe Dom Pedro, Scully argued that the failure could be put down to poor communications. Defending the colony's mainly Irish immigrants, he dismissed as unfair the accusations of laziness, arguing instead that, given the location of the colony, they were entirely helpless.[42]

By 1874, as diplomatic frictions between Britain and Brazil intensified over the plight of the immigrants, Scully became more defensive of the state colonies, claiming that settlers had "fair prospects before them". He dismissed those returning to England as "pseudo-martyrs of Brazilian wiles, immigrants of the least useful classes, in many cases the scourings of the poorhouses, whose ill conduct, turbulence, laziness and un-English incapacity for self-help strained to the uttermost the forbearance of the long-suffering Brazilian government and alienated all those who sought to aid their countrymen in their difficulties". The Brazilian government, Scully insisted, did what it could, but "the evil trades union virus" that had infected the immigrants was frustrating its good efforts.[43] That said, Scully also maintained that both Britain and Brazil still had much to gain if the immigrants could be encouraged to remain in the state colonies and if others could be persuaded to join them:

It is to the benefit of the English poor and English commerce that nuclei of British colonization should be successfully established in this country, in which life and property are safe, under constitutional monarchical institutions like those of England, from banditry and revolution. A few small prosperous nuclei will ere long induce spontaneous immigration in thousands to a country of settled and peaceful rule, whose delightful climate and many other natural advantages place it intrinsically one of the first in rank among the future homes of European immigration.[44]

But Scully's views shifted again, referring to state colonization as being "an insatiable quicksand which swallows up the public moneys". Even so, the blame was procedural, Scully maintaining that fine immigration policies had been undermined by their implementation.[45] He identified several factors for this failure, the causes less to do with the immigrants than the administration: the mal-selection of location, lands and immigrants; the absence of men properly qualified for responsible posts, willing to accept the arduous and badly paid senior administrative positions in a state colony; the discouragement of the colonists at the lack of provisions on their arrival, at the roadless wilderness in which immigrants were "dumped", at the sight of the lands reserved for them, frequently mere precipitous mountainsides. Scully also voiced sympathy for the Brazilian government which he believed was often deceived and cheated by its employees on the ground and he could understand its concern at the immense sums that had been spent on the state colonies with such slight results.[46]

As for Quintino Bocaiúva, Scully's Sociedade Internacional da Imigração colleague, he reluctantly accepted that it had been an error to set about recruiting immigrants in the United States, an approach which he admitted had been Utopian, artificial and doomed to failure. Nevertheless he insisted that the ideals behind the initiative were noble, that the affair had been patriotic in its motivation, carrying with it one overriding objective – "the good of the Fatherland and the greater future of the country".[47] The experience led Bocaiúva to believe that white Europeans could be extremely troublesome and that they were not necessarily hard working. It was far easier to point the finger of blame in the direction of the immigrants themselves – in this case falling back on the old stereotype of 'wild Irish' succumbing to the bottle and brawls – than to accept responsibility for a flawed scheme.[48]

Due to their limited resources, rural labourers were amongst those most vulnerable to exploitative or ill-conceived settlement schemes in places where no pre-existing social ties existed. The historian José Moya refers in a Spanish context to people recruited by government, private or semi-official as a "sudden gush", characterizing such movements as based on "artificiality", as opposed to "spontaneity". "In the absence of immigrant social networks, artificial efforts to ferment immigration can even have counterproductive effects", explains Moya pointing to the experience of Canary Islanders when semi-official agents of the Argentine government and private profiteers sought prospective emigrants in the 1830s. All the necessary conditions were apparently in place: a country with land aplenty eager for immigrants and overpopulated islands in the midst of a period of economic upheaval. The agents advanced passages to about two hundred families, on condition that the money was repaid. But unable to do so within the prescribed deadline and – crucially – lacking the support of previously established relatives or friends, the islanders had little alternative but to contract themselves to Argentine employers, effectively becoming indentured labourers and often enduring the most difficult of conditions. Not only were these individuals prevented by circumstances from establishing themselves as a nucleus, able later to call on friends and relatives to join them, but from this experience originated a negative image of Argentina in the Canary Islands. Moya observes that, "The importance of immigrant networks, after all, transcended direct sponsorship. The letters home, the bragging returnees, and the tales of successful sons generated a cultural capital in villages of origin that stimulated emigration". Rather than encouraging tales of success and triumph, the Canaries' scheme, argues Moya, had the opposite effect of keeping islanders away from Argentina. Despite the fact that the Canary Islands were one of Spain's main centres of emigration, few people from there were later to choose that country as a destination.[49]

Such 'artificial' and exploitative Latin American land settlement schemes regularly attracted British emigrants, but usually only the most desperate. Those lured typically included the poor, living on the fringes of urban society, large families from the countryside unable to accumulate savings for passage money to established destinations and even members of the middle class who had fallen on hard times or were simply attracted by the apparently easy terms being offered.[50] In the

Conclusion

immediate wake of the collapse of Colônia Príncipe Dom Pedro, and shortly after the settlement of the first British immigrants in Cananéia and Assunguy, the British government's Emigration Board warned that Brazilian immigration terms were far too generous to be believed and that the most desperate and gullible were liable to be easily tempted.[51] Although failure as farmers was by no means uncommon for English and Irish emigrants, what was particularly difficult about the Brazilian experience was that there were few alternative opportunities and an absence of pre-existing social ties available locally to fall back on. Furthermore, such work that was available – such as labouring on British-owned railways or domestic service with British families – was seen to pay little and offer conditions no better and often worse, than what had been left behind in England. Thus the only safety net available was to call on the assistance of British diplomatic representatives who, when pressed, would generally help distressed subjects to leave the country.

*   *   *

In many respects the experience of the Irish and English immigrants attracted to Brazilian state colonies between 1867 and 1873 mirrored those of other nationalities. While Príncipe Dom Pedro, Cananéia and Assunguy had their specific combinations of problems, these were similar to those of most other state colonies that had been established earlier. Remote locations and poor communications, land unsuited to European farming techniques, poor administrations and lack of resources all conspired against successful outcomes for immigrants, no matter how determined to succeed – as many certainly were. Many of the Irish and English immigrants were clearly completely unsuited to the rigours of pioneer settlement. But even some of these (or at least those who survived hunger, illness and other personal tragedies) might well have remained – if not actually prospered – had they not felt able to fall back on assistance from British diplomatic or consular representatives in Brazil. Príncipe Dom Pedro, Cananéia and Assunguy were the last attempts by the Brazilian government to create and manage so directly agricultural colonies and the Irish and English who were attracted to them amongst the last immigrants attracted to such ventures.

For Brazil, the failure to thrive emphasized that state colonization was an ill thought out policy, often poorly executed and usually under funded. But such an assessment of settlement attempts over a period of almost fifty years is not entirely fair: describing state colonization as a failure would ignore the fact that a quarter of a million immigrants went to Brazil between 1822 and 1870 (in particular between 1850 and 1870) of whom one-third were recruited on behalf of the Brazilian government.[52] While individual failures were many, these early arrivals contributed greatly to the future growth of Brazil's southern provinces and provided valuable lessons for immigrant-based land settlement policy.

In Britain, the Brazilian emigration episodes were an embarrassment for all concerned. Although numerous land settlement schemes in Latin America (including Brazil) continued over the next fifty or so years to attract British participants, no others recruited such numbers and most were independent initiatives with only indirect government involvement. However, as the nineteenth century progressed, Brazil remained a destination for Britons in search of prosperity. But few of these skilled workers, engineers, managers and businessmen regarded themselves as 'immigrants', instead travelling out on contracts to work for British companies, fully expecting to return 'home' in due course rather than having any intention of laying down permanent roots in an adoptive country.

Epilogue

VERY LITTLE OBVIOUS evidence is left in Brazil of the presence of the hundreds – perhaps as many as a few thousand – of English, Irish and Irish-American immigrants who were attracted as agricultural settlers between 1867 and 1873. In marked contrast to the experience of those pioneers, earlier and later waves of European (and also Japanese) immigration to Paraná, Santa Catarina, Rio Grande do Sul, São Paulo and Epírito Santo have left clear and present imprints. Due to a combination of factors – the higher number and concentration of the arrivals, external assistance and a determination to succeed, or some cases nowhere to move on to or even nowhere left to return to – these communities developed and some of them even prospered. As a consequence of isolation from 'mainstream' Luso-Brazilian society, in areas where particular nationalities or ethnic groups concentrated, cultural traditions were maintained. For example, today in rural areas of Santa Catarina and Rio Grande do Sul, one still finds communities where German dialects dominate, 140 years or more since the arrival of their immigrant ancestors.

In contrast, the fate and legacy of Brazil's English and Irish *colonos* are barely apparent. With each passing generation, only the faintest of impressions have remained of the British immigrants who settled in Brazilian agricultural colonies during the 1860s and 1870s. Nevertheless, slight traces of their presence can be found.

Colônia Príncipe Dom Pedro – Brusque

By 1871 all the surviving Irish and English immigrants had abandoned Colônia Príncipe Dom Pedro. Most of the former *colonos* returned to England or to the United States (or, in the case of some, went there for the first time). Others went to Argentina, a country just beginning to develop as a significant immigrant-receiving destination. A few of the *colonos* remained in Brazil, either finding work in the port of Santos or on railway construction elsewhere in the province of São Paulo, or agreeing to the Brazilian government's offers of being relocated to other state colonies, in particular to Assunguy. Neighbouring Colônia Brusque formally absorbed Príncipe Dom Pedro's territory and shortly after began to attract considerable numbers of Polish and Italian immigrants. During the twentieth century Brusque developed into a prosperous textile-producing city, although it is still occasionally afflicted, like other towns and cities in Santa Catarina's Itajaí-Mirim Valley, by floods.

Colônia Assunguy – Cêrro Azul

On 9 November 1887, 32 British male residents, who claimed to represent at least 165 men, women and children, of Cêrro Azul (the former Colônia Assunguy) assembled at the town hall to draw up a statement describing local conditions. After almost twenty years since the arrival of the first British immigrants, little seemed to have changed. There was still no wagon road linking Curitiba with Cêrro Azul and the mule track had, if anything, deteriorated ("bridges being very dangerous, many broken down altogether, risking life and limb, also animals") since the withdrawal of all central government support some years earlier. There was still no bridge across the Ribeira River linking the areas where most of the British were settled, the raft being only usable when the river was low. Two years earlier the raft had been swept away and for twelve months they were dependent on canoes and swimming animals across the river. The British *colonos* complained that although they were being taxed on the animals and sugar that they managed to sell, no money was being spent on bridges, roads or local services and there was still no resident Protestant minister, while the nearest medical doctor was in Curitiba. To survive, a great deal of effort was required of farmers as much of the productive land was on hillsides. As a consequence, the best off were those with large families whose members could contribute labour. Agricultural production was still frustrated by the fact that produce taken to market in Curitiba would lose half of its value through transportation costs, and there was no point in carrying corn, the crop best suited to local conditions. All in all, the *colonos* felt that most of

Epilogue

them were only making a bare living, even by the meagre standards of Brazilian small-scale agricultural production.[1]

But a British community, of sorts, had survived although cohesion was now based on occasional meetings called to draw up lists of complaints to put before consular representatives in Santos or Rio. As the years passed, however, these officials were more powerless than ever to intervene on behalf of British subjects as the Ministry of Agriculture had severed its ties with (and suspended financial support for) Assunguy in 1882 when the new *município* of Cêrro Azul was created. Conditions in Assunguy had always been harsh, but a hard core of immigrants had persevered, with some of the men (who had gone to work in Curitiba to support their families back on their farms) opting to return to the colony. Although transportation links with Curitiba failed to improve, reasonable proximity to the steadily expanding state capital at least offered an employment safety net where they could count on the support of a network of friends and extended family. By 1890 the community had stabilized, with the Germans and Irish considered to be the best off of Cêrro Azul's ethnically diverse population.[2]

Apart from English and Irish immigrants and their descendants, there remained in Cêrro Azul people of French, German, Swiss and

A Chamberlain-family wedding gathering, Cêrro Azul, c. 1927.
(Margarida Chamberlain)

Three generations of the Chamberlain family.

Italian origin, as well as many of Luso-Brazilian background. Given the mixed nature of Cêrro Azul's population and the territorial spread of *colonos* of different ethnic backgrounds, it is hardly surprising that mixed marriages soon became the norm, with Portuguese being the common language within families. Today the names Blane, Butcher, Chamberlain and Fitz (from Fitzgerald) are some of the Irish and English surnames still to be found in Cêrro Azul, but none of their bearers have more than the faintest knowledge (or interest) in their immigrant origins.[3] For example, Carlito Chamberlain, born in 1931, only knows that his grandfather, Alberto [Albert] Chamberlain (who died in 1907, long before Carlito's birth), came from "Europe" with his father, Shadrach. Luiza da Conceição Blane, born in 1921, recalls her father speaking with her grandparents in a "foreign language" and that they had come to what was Assunguy from Santa Catarina due to problems of some kind that they had experienced there. Ernesto Fitz, born in 1914, Dona Luiza's husband, knows that his grandfather, João [John] Fitz[gerald], had come to Brazil as a child.

Instead, preoccupations of Cêrro Azul's present-day population are naturally far more practical. Despite being located just over one hundred kilometres from Curitiba, one of Brazil's largest and most dynamic cities, economic development has made slow progress in Cêrro Azul.

Epilogue

Luiza da Conceição Blane and Ernesto Fitz.

This has primarily been due to continued communications limitations, with the *município* being one of the few parts of Paraná that is not directly linked to the state's otherwise excellent highway network by a paved road. Although the road connecting Curitiba to the town of Rio Branco do Sul (formerly the hamlet of Votuverava, the usual midway rest point for mule trains) is of a good quality, it then dramatically deteriorates as it winds its way along mountain passes, the potholed and rutted dirt road equally treacherous in dry weather as in wet. Asphalt is, at long last, being laid along this remaining stretch of the road. This improvement was held back in recent decades not so much due to cost or engineering difficulties but because it was feared that improved communications would encourage Cêrro Azul to expand and consequently risk polluting the important local water reserves which the city of Curitiba might require if those of the Serra do Mar cease to be sufficient.

Cêrro Azul's administrative and commercial centre retains the general look and feel of a village rather than that of a town. At its heart is the Praça Monsenhor Celso, originally the Pateo da Colônia, a large area of former marshland once considered a health hazard but which has long since been filled in and grassed over for use as a recreation ground. At one end of the Praça is the Catholic church, while bordering it are rows of simple one and two storey buildings (a few of which date back to the 1860s) that have various commercial and administrative functions.

Most of Cêrro Azul's population of nineteen thousand live widely dispersed in the *município*'s rural interior. There, almost all the inhabitants now have ready access to electricity, schools, medical care, a limited bus service, chapels for Catholic worship and, as is typical of throughout Brazil, a choice of Protestant evangelical missions. Leaving the immediate vicinity of the urban centre, roads are still appalling, but at least bridges cross the Ribeira and Jaguatirca rivers and there are also some strong, if rickety-looking, wooden structures that cross the Turvo to give access to what otherwise would be entirely isolated farmsteads. Most of the rural inhabitants eke out livings on plots measuring at most just a few hectares. With productive land often being of a bewildering gradient, the *colonos*' properties require as much hard work as they ever have to cultivate successfully. Beans and corn remain the staples, still mainly produced for domestic consumption, but at last a viable commercial crop has been found. Excellent quality citrus fruit – tangerines, but also limes and oranges – are produced on the frost-free lower ground, and this is taken by truck to Curitiba for processing or onward distribution.[4]

Colônia Cananéia

The great majority of Cananéia's Irish and English immigrants abandoned the colony as soon as they possibly could, with those who persevered being mainly from amongst the 1868 arrivals. Nevertheless, a few lingered, tied down by growing families and the properties into which

Epilogue

so much effort had been poured. One of the *colonos* who remained the longest was John Mitchell who appealed in 1881 to the British legation in Rio for five hundred milreis (around £54) to enable himself, his wife and three adult children to travel to Canada. He said that the family had fully paid for their five hundred acre property, but that it was impossible either to sell or to use as collateral for a loan.[5] A few years later James Davies, one of the Birmingham recruits who had travelled to Brazil aboard the *Florence Chipman*, explained simply why he remained in Cananéia when most other Britons had long given up: "my family increased in the colony so much that I could never rais a nough mony to lave the colony i have six sons and two daughters sir" [*sic*].[6]

Despite being far from any real centre of population, even by the time the English and Irish inhabitants had dwindled to just a few families, the colony was not entirely forgotten by the outside world. At least through the 1880s, British consular staff, as well as family members, themselves struggling in Santos or elsewhere, sent occasional handouts.[7] Although consular visits were impractical, in October 1888 the HMS *Rifleman*, a British naval vessel, called, anchoring off the port of the colony. The *Rifleman*'s commander and surgeon set off by foot, stopping along the way at the homes of a German family and a British man, his Swiss wife and their three children. Some twenty kilometres from the port they reached the property of Bridget Davies, whose husband, James, had returned to Birmingham two years earlier, where he died soon after.[8] Lieutenant Commander May left the following report:

> Mrs Davies lives with her family, viz: 2 daughters and 5 sons, the eldest daughter a woman of about 33 is a cripple, and the other daughter aged 17 is a fine well built young woman. The eldest son is 25, the youngest 4 years old. Their house is built in the native style, plenty of pigs and fowls. They also own 3 horses, cows won't live on their land, they have often tried, but soon die, something to do with the grass they think. What surprised me the most was the excellent English they all spoke. They appear very cheerful, to have a hard struggle to get on. They have to work hard to grow sufficient food for their own consumption, as the weeds and bushes grow at such an alarming pace, that they have to be constantly clearing the ground. They complain of the state of the roads. For 1000 milreis they would undertake to put them in sufficiently good order for

packhorses. A sister in law of Mrs Davies lives about a mile from there. She is married to an Irishman called Murray. The eldest son assists his father who is one of the original settlers. There are several Brazilians in the colony but they won't work more than they can possibly help, and don't care whether the roads are in repair or not, as they sell nothing.

I asked Mrs Davies if she would like to go home, she said no she had been out there so many years. I savvy the sons would be glad to leave if they could find their way to any other employment.

I call it a miserable place, everlasting jungle and forest, an almost sultrie absence of animal life, except poisonous snakes, huge toads, and large horse flies, which attack both man and beast.[9]

As evening approached, horses were saddled and one of Mrs Davies' sons escorted the visitors back to the *Rifleman*, returning home the following day laden with medicines, books, old clothes, boots, hats, biscuits and other items that the crew had donated.[10]

Today, with a population of around twelve thousand, the town and rural hinterland of Cananéia remain a backwater, the region being one of the poorest parts of São Paulo, Brazil's wealthiest state. Thanks to Cananéia's isolation and limited economic development, there is still a considerable amount of primary vegetation, with much of the region now classified as an environmentally protected area, the forest and mangrove being some of the very little remaining of the *Mata Atlântica* (Atlantic forest) that originally covered much of the coast of Brazil. In addition to fishing and agriculture, Cananéia has developed in recent years as a modest tourist centre: the town itself has some recently restored, extremely simple but pretty, colonial-era buildings, while the nearby beaches and islands have remained largely unspoiled due to the area's poor roads and the distance from large centres of population.

In dry weather the area of what was once Colônia Cananéia – now marked on local maps and signposts as the "Ex-Colônia" – can be reached from the town in about an hour. The dirt-and-gravel road is potholed along many of its stretches but it is usually easily passable by car. The route winds its way inland from the mangrove-fringed shoreline where the colony's 'port' was located (now a ferryboat crossing point linking the mainland with the island of Cananéia) and successively passes through scrubland, secondary forest, banana plantations and cattle pasture. Several families live in the Ex-Colônia, some of whom are said to be descendants of men and women who created a *quilombo*, a community of escaped slaves, on land vacated by the European immigrants. The sole remaining residents of British origin are brothers David and João Davies, remarkably sprightly octogenarians, who proudly explain that they are grandsons of William Davies, a son of James and Bridget Davies. William, they know from a photocopy in their possession of an old British passport, was born in Birmingham on 22 June 1859 – making him nine years old when he arrived in Cananéia – and they also know that he was married to a German woman of whom all that is still remembered is her first name, Anna. David and João talk animatedly about their grandfather and great-grandparents, explaining that the family's strength and determination to overcome hardships lay with Bridget, condemning James for deserting both her and their children when he returned to England. They climb over fences and leap across ditches to reach the cemetery where Bridget and many of the other immigrants were finally laid to rest, but the burial mounds are overgrown and any names that might have once been inscribed on the wooden or iron crosses have long-since faded away.

João and David Davies.

Almost all of the Ex-Colônia's limited expanse of flat or gently sloping land is now given over to pasture for water buffalo; the property once mostly owned by the Davies family was sold off little-by-little to a big city corporation with vast land interests elsewhere in Brazil, to help pay for the Davies' children's education and make their lives a bit more comfortable in old age. Although buffalo appear well suited to the hot and damp conditions of this part of coastal Brazil, raising such cattle is unlikely to be sustainable on a long-term basis. The main problem is the way the animals excavate the soil, digging to create ponds and thereby accelerating the erosion process of the pasture that is probably irreversible.

Electricity reached the Ex-Colônia in 1983 and access to family members in the nearby towns of Jacupiranga, Pariquera-Açu, Registro, Iguape (as well as further afield) is comparatively easy, if uncomfortable, by road. But life remains extremely simple. David and João Davies and their wives tend their remaining small portions of farmland, growing corn, beans, coffee, fruit and other crops and raising chickens and a few pigs, making themselves virtually self-sufficient. The family, like pretty well all the Ex-Colônia's inhabitants, are devout Presbyterians – a Protestant denomination introduced from Rio de Janeiro by North American and Brazilian missionaries a century or so ago – with regular Bible readings and occasional services given by an itinerant pastor conducted in a chapel located on a hillock in the midst of the valley. David and João Davies are especially unusual: they knew personally their immigrant grandfather and are last surviving descendants to have had that contact. But within a few years such a direct connection with the past will be entirely broken.[11]

Appendices

Appendix 1

A petition to Pope Pius IX

Addressed to Pope Pius IX, the petition reproduced below was published in *The Rev. G. Montgomery's Register* (vol. 1, no. VI, 19 October 1867). In his introduction, Montgomery explained that he only permitted to sign the petition men whose "upright and unblemished character" he felt able to vouch for personally and with confidence. Montgomery claimed that he read the entire text to all who were invited to sign the petition, explaining the contents to every individual until he was certain that it was fully understood (most of those committing their names to the petition would have been illiterate). All told, the 96 signatories – "all heads of families" – would have represented several hundred potential emigrants, although it is unknown how many in fact proceeded to Brazil.

The Humble Memorial & Petition of Certain Irish Catholics who Sojourn in England, to Our Apostolic Lord, His Holiness Pope Pius the Ninth.

We are fathers, heads of families, natives of Ireland; who, pressed by poverty, have left the land of our birth, and sojourn in England. We support ourselves by manual labour, for most part of the rudest sort, and depend for employment chiefly on the great manufacturing industry of this country. When the trade of England languishes, there is little or no need of our services, and we are frequently altogether deprived of employment for many weeks together. In short, our temporal condition is entirely at the disposal of persons who have no relation to us but that of employers, who, so far as we are concerned, using their money only to make more money, hire us to work, or dismiss us to idleness, as their interests require.

When we are dismissed or suspended from employment, we must leave our families, and wander about "looking for work"; or live in a half-naked, half-famished state, getting miserably in debt; or, breaking up such homes as we have, we must seek shelter in *Poorhouses*.

If we travel about looking for work, we are in danger of departing far from our neighbourhood of priest and altar, and thus of seeming, like Cain, *"to go out from the face of the Lord, and dwell as fugitives on the earth."* If we enter the poor-houses, we go to imprisonment, to forceful companionship with persons not Catholics, who may be hateful to us; we must submit to the yoke of rules which are oppressive, because they were not framed with our consent; and too often are not administered with kindness, but are, in some instances, repugnant to the laws of the Catholic Church; and are, at best, but regulations for the orderly dispensing of relief, grudgingly and of necessity given to the poor. How dreadful the thought that we might die in these places, or that our young children may be immured to them to grow up listless, faithless parish paupers, having in after-life to struggle for a place in the lowest grade of the social scale, though we had hoped to rear our children to be in all respects better off in this life than we ourselves have been!

We are told by some persons that we are improvident, and that in prosperous seasons we might lay up something for the time of distress. Some of us do indeed strive to provide for periods of want, by reserving a portion of our earning as contributions to a fund out of which we may receive some aid when sickness or any bodily accident befalls us. But we cannot do this and provide for the times when we are out of work; and many of us have helpless children to support, and aged parents, and other necessitous relatives, whom we must aid. It is but seldom that we commit any wasteful excess; and if we are not duly economical, perhaps it is because we have not had the good fortune to be taught how to be so. Do our best, we suffer that extreme and compulsory poverty against which we are taught by the Holy Scriptures to pray.

To the evil of our extreme poverty there is added this other, – that we are strangers in the land, disliked by the people amongst whom we live, because of our nationality, because of our religion, and because we are in competition with them for employment. During the hours of our work, we have to associate with persons who assail us with blasphemy against the most sacred doctrines of the Catholic religion, with defamation of the clergy and female religious persons, and with obscene discourse.

Appendix 1

We know also that those around us attribute our poverty, our faults, our follies, and our crimes to the influence of the Catholic faith. This terrible storm of persecution, of calumny, is sufficient to overwhelm persons more steadfast than we are; and we tremble when we think of the effect which it has on our children. At all times this tempest is felt by the poor Irish in England, but just now – excited by the fraud and malice of certain fanatics and apostates – it rages with fury against us. So we, who are sociably inclined, are forced to keep aloof as much as possible from the people of the land, lest we be terrified or seduced from our attachment to the faith.

We must complain, too, that the conditions under which we live, as mere labourers in the places where we get employment, and only as a minority of the general population, prevent us from separating ourselves and our children from the neighbourhood and companionship of certain Catholics – our countrymen too – who openly and constantly violate divine and human laws; persons who neglect all religious duties, and abandon themselves to drunkenness, and the squalor and shameful habits consequent upon irreligion and intemperance. In the places where most of us reside, there are many such Catholics; but there is not one Catholic employer, nor one who occupies high social position, – not one to afford us patronage.

Another evil in the natural order which afflicts us is, that our children, who are born and grow up in England, must grow up without patriotism; for we cannot teach them to love a country which has departed from the Church and is hostile to it, and which used its power for many ages to oppress our native land, and to extinguish in it the light of faith. We continually hear our fellow-Catholics – the English – proclaiming their love of their country, and their great loyalty to its Government, though that Government is alien to the Church, and treats the Vicar of God with contempt; and account it hard that we should be called on and expect to think and speak like English Catholics. The mere fact that we came to this country to labour for a living, and that poverty compels us to remain in it, is not sufficient to make us love it, nor cause us to teach our children to love it.

We do *not* complain that the cost to us of the maintenance of our religion is, in proportion to our means, very considerable; but we complain that, except in our churches, – which we cannot frequent daily, – we scarcely see or hear anything to remind us that that the Catholic religion exists;

and we complain that we, an illiterate people, have not the moral and religious support which, in the conversation and usages of every-day life, is afforded in other lands of similar social conditions to our own. By continual contradictions and blasphemies and ridicule directed against Catholic doctrines and practices, we are indeed reminded of the religion we profess. But if these things do not detach us, natives of Ireland, from the Church, they tend not a little to cause the crowning evil of our present state; that is, a too well-grounded fear that our children, or our children's children, will apostatize. We have seen many children, born in England of Irish-Catholic parents, make shipwreck of faith and morality; and some of our clergy confidently assert that the children of Irish in Great Britain who fall away from the faith far exceed in number the natives of the land who are converted to the Catholic Church.

We are painfully conscious of the evils which afflict us; we groan and look up to God; we groan as a people persecuted, – persecuted by the pressure of poverty, which has made us exiles, which keeps us always strangers, and often wanderers, in a land where we are degraded, insulted, calumniated, importunately tempted to vice and heresy, plunged in tribulation, *"pressed out of measure above our strength, so that we are weary even of life."*

What shall we do? Some of our friends tell us to be pious, to be patient, and hope for better times in this land. But our sense of the evils which encompass us is too keen to allow us to be tranquil. Speaking with others similarly placed as ourselves, we say, – doubtless the prelates and pastors and missionary clergy of the Church in England do for us what they can, labouring for us zealously, patiently, and with tender compassion. But we cannot help hearing that these, our loved and honoured friends, say, or hint, that we cannot remain in the land, and refuse to become English; that we are a wayward and troublesome people; that the number and greatness of our necessities far exceed their means to relieve them adequately; and that we seem to be a doomed race, that rapidly tends to extinction. We cannot help hearing that these things are said of us by our friends, and we hang our heads in shame and sorrow; but we despair not. We refuse to be absorbed in the English nation; we shrink from the prospect of the extinction of our race; we shudder with horror at the idea of our children becoming apostates, and deriding the faith and the birth-place of their fathers; and we look with dread and dismay upon the land in which an enormous proportion of our

Appendix 1

people, old and young, are numbered as paupers and criminals.

We confess that all that has come upon us has happened by the permission and the just judgement of God. We hear our divine Saviour saying, "*When they persecute you in one state, flee ye to another,*" and we look whither we may flee to obey this precept, follow this counsel, or avail ourselves of this permission – whichever it may be in our regard. To the United States of America many of our kindred and friends have recently gone, and there they are, in comparison with us, in temporal prosperity, and religiously they are better circumstanced than we are. We look wistfully after them. We cannot follow them; we contemplate our misery sojourning here, but we despair not of ourselves. We have cried to the divine Jesus for mercy, to our Lady for help, to the Vicar of God for his blessing; and the brightness of hope has illuminated our path. Our hope is that we may be received into the empire of Brazil, where such persons as we are, are wanted and would be welcomed; there to find a home – a dwelling-place whence we cannot be expelled at the mere will of others – and means to be on our own lands constantly employed, and no longer the sport of the fluctuations of trade; there to find a people the vast majority of whom are Catholics – a sovereign who recognizes the divine Jesus in the Sacrament of the altar, and who bows before mysteries which are here made the butt of the unbelieving of scoff and ridicule; there to find a Government Catholic by law.

Signed,
Thomas MacGeoghegan, John Shannahan, William Farrell, Miles Kirby, John Kirby, John Connolly, John Gallagher, Patrick Kavanah, Martin Morrin, Patrick Swift, Edward Baxter, Mathew Burns, John Hopkins, Michael Lee, Thomas Walsh, Timothy Holahan, William Fitzgerald, John A. Slater, Patrick Madican, Stephen Collins, John Brennan, Patrick Lyndon, Peter Costello, John Grady, Patrick Biggins, Patrick Joyce, Martin M'Tighe, Michael Jordon, Patrick Lynch, Thomas Clark, John Driscoll, Michael Mylott, Patrick Butler, Michael Walsh, Hugh Brady, David Harris, John Joyce, Martin Cohen, John Brannigan, Patick O'Brien, John Byrne, Patrick Cuningham, Thomas Connell, John Walsh, Patrick Connelly, Patrick Kenny, Michael Corless, John M'Donald, Martin Flaherty, John Carroll, John Cavanah, Timothy Kelly, Patrick Hegarty, James Craddock, James O'Neal, Thomas Cavanah, Jerimiah Monghan, Roger Loyden, Martin Moran, John M'Grath, Patrick Shereden, Patrick Grelly, Patrick Hopkins, Patrick Joyce, Anthony Nolan, Michael M'Farlin,

Thomas Walsh, Michael Colleran, Thomas Morrin, Patrick Walsh, Michael Gilmore, Patrick M'Farlin, Patrick Cregan, Patrick Geraghty, John Flaherty, Patrick Jennings, John Cuningham, John Naughton, Michael Varley, Martin Tracy, John M'Donough, John Flynn, John Brannigan, John Burns, James Grimes, John Cavanagh, Thomas Ronan, John Halloran, Luke Joyce, William Feehan, Patrick Jennings, Martin Collins, James Shannahan, Patrick Kelly, Charles Connor.

Appendix 2

1898: A final appeal for assistance

Over the course of thirty years, British diplomats and consular officers in Rio de Janeiro, Santos and elsewhere in Brazil received a steady flow of appeals for help from residents, or former residents, of Príncipe Dom Pedro, Cananéia and Assunguy. Originally from Westmeath, Ireland, but for some time a resident of Birmingham, Bridget Davies arrived in Brazil in 1868 with her English husband, James, and their children, on the *Florence Chipman*; the family was amongst the first of the British to settle in Cananéia. James returned alone to Birmingham in 1886, but Bridget remained in Cananéia through her old age, with some of their descendants still to be found in the 'Ex-Côlonia'. The following appears to be Bridget Davies' last letter to British diplomatic representatives in Rio de Janeiro – as well as being quite likely the last appeal for assistance from a British resident of one of the Brazilian former state colonies.

Bridget Davies (Cananéia) to British Minister (Rio de Janeiro), 25 September 1898 [1]

Colonia de Cananea province san Paulo September the 25 1898 to his excellence the English Minester in riodejanerio sir you will plase to excuse me trespassing on your valuable time but i am in truble and i have applied to the consilate of santos and san paulo but the have not answered me sir you will plase to excuse me i do not know your name, i am a English colunist living in the colonia of Cananea my husband James Davies came here in 68 he died in 87 laveing me with a large family of children we have suffered much hardship trying to make a home and exspecting to have the Benefits of the colonia for our children i sent my youngest son walter to study for a school master, as there was a boys and a girls school in the colonia he study two years while he was studying a nother man in cananea took the place my son fineshed his studes this man said he would sel him the place for 400 milreis and he could pas in cananea and if he went to sanpaulo it would cost more, so my son told me, my son had no mony so i sold my coffy my whole years work, and my other childrens, to raise the mony my son was examined and sent his papers to sanpaulo the first of august last year the camere wrote for him he failed up til Chismas the hid master said it will come in the knew year in january there came in pecthore a german and my son told him about his papers in sted of that the first journal from sa Paulo gave notice no more boys school in the colonia of canaea sir i hope you will have patituns i want to explain the truth the school mistress has very few girls never more than eight and she offered to lern the boys and as well sir i unble ask you to inter-sais for my son with the government of san Paulo i belive the would give him the school and if the knew the truth we have lived 30 years in this colonia and tride to promote the colonia and I believe if you would intersaide for us my son is a very good young man and a good scholar in Brazilian his own language is English a few words of French there are eight English famellys here all good respectable peopol and it is hard that we cannot have one of our children to be a school master we have not a paper nor a book of any kind we would be truly glad of the Books or papers that you have Red 12 years ago Mr Gough sent me some books and papers and medecin a good English gentel man my son walter Davies he is 20 next month and if you did not get him a place as clerk in any good firm i beg you to have paitions with my bad writing i am old and anxcious for my son, sir plase to favor me with a answer so I conclude my scribble remaining your

    my address   umble servant
    the wido Davies   the wido
    Colonia the Cananea   Davies

Appendix 3

British immigrants in Príncipe Dom Pedro, Cananéia and Assunguy

Passenger lists for ships bound for Brazil from Britain in the late 1860s and early 1870s have not survived in either the National Archives (Kew) or in archival collections in Liverpool or other British ports. Passport records for the period are also lacking owing to the fact that emigrants destined for government agricultural colonies travelled on passports issued by Brazilian consuls.

There is also an absence of official records detailing immigrants arriving in Brazil during this period. Searches have yielded no relevant documents held by the Arquivo Nacional, Arquivo Histórico Diplomático (Itamaraty), Biblioteca Nacional and Arquivo da Marina collections, or by the city archives of Rio de Janeiro and Paranaguá. In addition, no records appear to survive from the Casa de Saúde, registers only existing from 1879 when the Ilha das Flores was inaugurated as Rio's immigrants' hostel.

The following table has been compiled from a variety of alternative sources. Foreign Office files include details of many emigrants, with names also found in parliamentary reports, Brazilian and British newspaper articles, provincial government debt ledgers and family records. Although numbering over 1,200 individuals, inevitably this is only a partial listing of the British *colonos*, in particular lacking details of women and children.

Abbreviations

b. = born
d. = died
? = uncertain
+ = remained at least to that year

ABT = *The Anglo-Brazilian Times*
APP = Arquivo Público do Paraná, Códices
Assunguy = Great Britain, *Emigration to Brazil. Report on the Colony of Assunguy (1875)*
Cananea = Great Britain, *Emigration to Brazil. Report on the Colony of Cananea (1875)*
FO = National Archives, Foreign Office
LUC = *The Labourers' Union Chronicle*
PP 1873 = Great Britain, *Reports Respecting the Condition of British Emigrants in Brazil* (1873)
PP 1874 = Great Britain, *Report Respecting the Condition of British Emigrants in Brazil* (1874)
RLSC = *The Royal Leamington Spa Courier and Warwickshire Standard*
Tigar = Albert Tigar manuscript (Tigar family)
Times = *The Times*
WGC = *The Wilts and Gloucestershire Standard*

Appendix 3

British immigrants in Príncipe Dom Pedro, Cananéia and Assunguy

NAME (age on arrival)	Occupation	Marital status	Family composition (age on arrival)	Place of origin	Arrival year	Colony	Depart year	Last known destination	Sources
Andrew Adams					1868	Assunguy			APP 0387 p. 178
James Adams (27)	bricklayer	single			1872/3	Curitiba/ Assunguy?	1873	Rio/?	FO128/98/506
Edward Allen					1868	Assunguy	1869+		FO128/92/260; APP 0387 p. 167
John Andrews	labourer	married	7 children	Dorset	1872/3	Assunguy	1874+		*Assunguy*
J. William Apperley		Anne			1872	Assunguy	1873	England	*LUC* 4/1/1873; *Times* 15/10/73; FO128/100/169
Elijah Arnold (41)	sawyer	Mary Ann (41)	Mary Matilda (8)		1872/3	Curitiba/ Assunguy	1873	Rio/?	FO128/98/506
George Arnold	agricultural labourer	married	"children" (of whom 2 d.)	Sarsden, Oxfordshire	1872/3	Assunguy [Bariguy]	1873	England	*Times* 17/6/73; WGS 21/6/73
James Austen						Assunguy	1874+		APP 0387 p. 148
Henry Bailey		married				Cananéia	1873+		FO128/100/198
Ammon Barker	labourer	married	5 children			Assunguy	1888+		FO128/109/463; FO128/155
Edward Barnard		single			1873	Cananéia	1873	Rio/?	PP 1874; FO128/101/349

229

British immigrants in Príncipe Dom Pedro, Cananéia and Assunguy

NAME (age on arrival)	Occupation	Marital status	Family composition (age on arrival)	Place of origin	Arrival year	Colony	Depart year	Last known destination	Sources
John Barton	carpenter	married	5 children (some b. Brazil?)		1868	Cananéia	1874+		ABT 11/1868; Cananea
Thomas Bayliss (d. Rio 1873)	railway worker/ blacksmith	Louisa		Gloucestershire	1872	Cananéia	1873?	England	FO128/101/300; PP, 1874
George Beer					1873	Cananéia	1873	Rio/?	PP 1874
Mr Bench		single				Assunguy	1874+		Assunguy
Henry Benson				Warwickshire	1872		1873	England	PP 1874; Times 17/10/73
John Bignell (27)		Anne (28)	Frederick (9), Eliza (7), Charles (5), Emma (3)		1872	Cananéia	1873	England	FO128/100/173; Times 15/10/73
Charles Birmingham		married	3 children	Gloucestershire	1872	Cananéia	1873	England	Times, 17/10/73
Jane Bishop		single				Assunguy	1874+		Assunguy
Alexander Blane						Assunguy	1874+		APP 0453 p. 28
George Stewart Boddy	agricultural labourer	married	7 children (some b. Brazil?)		1868	Assunguy	1890+		FO128/155; Assunguy; FO128/167; APP 0387 p. 139
William Boldwin					1868	Assunguy			APP 0387 p. 189

Appendix 3

British immigrants in Príncipe Dom Pedro, Cananéia and Assunguy – *continued*

NAME (age on arrival)	Occupation	Marital status	Family composition (age on arrival)	Place of origin	Arrival year	Colony	Depart year	Last known destination	Sources
Clara Bond (20) (daughter of George Bond)	single	Lucy	Annie Bertha (4 months)	Cirencester, Gloustershire	1873	Assunguy	1874	Randwick, Gloustershire	FO128/98/514; FO128/104/143; FO128/104/167; FO128/105/87; APP 0382 p. 150; family records
George Bond (d. nr Curitiba, May? 1873)	blacksmith	arrived as widower	Charles (13), Annie (11), Emma (9), Ernest (7), Alfred (3) – + Clara Bond (above)	Cirencester, Gloustershire	1873	[Assunguy]	1873	Curitiba	FO128/98/51; FO128/101/187
George Thomas Bouckridge (?)					1869?	Assunguy	1870?		APP 0387 p. 134
Henry Boryer	carpenter	married	2 children	Gloucestershire	1873	Assunguy	1888+		FO128/155; *Assunguy*
James Boryer (41)	carpenter	married	2 children		1872/3	Assunguy	1874+		FO128/98/506; *Assunguy*
William Boucher		married	1 child	Gloucestershire	1873	Cananéia	1873	England	PP 1874; *Times*, 17/10/73
Patrick Bowen		Margaret	6 children		1868	Príncipe Dom Pedro	1869	Rio/?	FO128/92/96
Mr Bradley					1867/8	Príncipe Dom Pedro	1870	Cananéia	FO128/95/186

231

British immigrants in Príncipe Dom Pedro, Cananéia and Assunguy – *continued*

NAME (age on arrival)	Occupation	Marital status	Family composition (age on arrival)	Place of origin	Arrival year	Colony	Depart year	Last known destination	Sources
Hugh Brady (37)		Mary (31)	Peter (11), Margaret (8)		1867/8	Príncipe Dom Pedro			FO128/94/243
William Bragg (21)					1872/3	Assunguy (Curitiba)	1874	England	FO128/104/167
Edward Brine						Assunguy			APP 0382 p. 18
Isaac Brain					1872	Cananéia			*Times*, 24/12/72
George Brine	agricultural labourer	married	5 children (some b. Brazil)	Winfrith, Dorset	1872	Assunguy	1888+		*LUC* 31/5/1873; FO128/155; *Assunguy*; APP 0382 p. 18
John Brine						Assunguy	1877+		APP 0453 p. 357
Manuel Brine	agricultural labourer	married				Assunguy	1888+		FO128/155
William Britten (or Briter?)					1868	Assunguy	1869+		FO128/92/260; APP 0387 p. 161
William Brown				Shotteswell, Oxfordshire	1872	Cananéia			*RLSC* 23/11/72
Henry Budding						Assunguy			APP 0387 p. 145
Frederick Burton	labourer/ miner		Catherine 7 children (some b. Brazil)	Staffordshire	1867	Assunguy	1888+		FO128/155; *Assunguy*; APP 0387 p. 115

Appendix 3

British immigrants in Príncipe Dom Pedro, Cananéia and Assunguy – *continued*

NAME (age on arrival)	Occupation	Marital status	Family composition (age on arrival)	Place of origin	Arrival year	Colony	Depart year	Last known destination	Sources
George Butcher	stonemason	married	5 children		1868	Assunguy	1888+		FO128/92/260; APP 0382 p.124; APP 0453 p. 194; APP 0387 p. 175; FO128/155
Tobias Butler (46)		Bridget (38)	James (18), John (15), Mary (10), Elizabeth (8), William (6), Anne (2)	"Irish"	1868	Príncipe Dom Pedro	1870		FO128/94/243; FO128/94/375; FO128/95/141
George Caddy (25)		Anna (25)	Thomas (2)				1873	England	*Times* 15/10/73; FO128/100/169
Mr Callopy ?		married	6 children	New York	1867	Príncipe Dom Pedro			FO128/94/375
John Cameron						Cananéia	1878+		FO128/116
Edward Capel		married	3 children	Oxfordshire	1872	Cananéia	1873+	Mendes, RJ/ England	PP 1874
Bernard Carroll (23)		Catherine (25)			1867/8	Príncipe Dom Pedro	1870+		FO128/94/244
John Carter		married	Emma (10), John (3) plus 1 d. Cananéia	Oxfordshire?	1872	Cananéia	1873	England	FO128/101/303; FO128/100/169; PP 1874; *Times* 15/10/73

British immigrants in Príncipe Dom Pedro, Cananéia and Assunguy – *continued*

NAME (age on arrival)	Occupation	Marital status	Family composition (age on arrival)	Place of origin	Arrival year	Colony	Depart year	Last known destination	Sources
Shadrach Chamberlain (28)	agricultural labourer	Mary	Edward (5), Sarah (6), Lizy [Eliza?], William (plus 5 b. Brazil)	North Wootton, Somerset	1872	Assunguy	1888+		FO128/106/339; FO128/155; *Assunguy*; family records
Henry Chester (18)					1867/8	Príncipe Dom Pedro			FO128/94/244
William Claridge					1868	Assunguy	1870+		FO128/92/305; FO128/92/261; FO128/92/304; APP 0387 p. 158
Martin Cohen (45)		Margaret (45)	Mary (18)	"Irish"	1867/8	Príncipe Dom Pedro	1870		FO128/94/243; FO128/94/376; FO128/95/141
Charles Coleman		married				Cananéia	1873+		FO128/100/198
Jeremiah Collins					1868	Cananéia	1868+		*ABT* 7/11/68
Thomas Collin					1868	Assunguy	1869+		FO128/92/260; APP 0387 p. 183
Patrick Collopy ?				New York	1867	Príncipe Dom Pedro			FO128/94/376
William Collopy ?				New York	1867	Príncipe Dom Pedro			FO128/94/376

Appendix 3

British immigrants in Príncipe Dom Pedro, Cananéia and Assunguy

NAME (age on arrival)	Occupation	Marital status	Family composition (age on arrival)	Place of origin	Arrival year	Colony	Depart year	Last known destination	Sources
William Connell (24)		Bridget (22)	Patrick (b. Brazil 1868)	"Irish"	1867/8	Príncipe Dom Pedro	1870		FO128/94/376; FO128/95/141
John Connolly				Wednesbury	1868	Príncipe Dom Pedro			*Universal News* 29/2/68
Robert Cook		married d.	4 children				1874	England	FO128/102/294
Thomas Arthur Cooke	railway worker			Leamington, Warwickshire	1872	Cananéia	1873+	England	FO128/101/150; PP 1874
Mr Cooper		married	3 children			Assunguy	1874+		*Assunguy*
James Cooper (33)		Sarah (34)	Thomas (12), Edward (9), Rose (7), Eliza (5), Laura (2)		1872		1872	England	*Times* 15/10/73; FO128/100/174
William Cornell (23)		Bridget (22)	(1 b. Brazil)		1867/8	Príncipe Dom Pedro			FO128/94/243
James Costigan d. Rio 1869		married	3 children		1868	Príncipe Dom Pedro	1869	Rio/?	FO128/92/95
Mr Cousins		married	6 children (some b. Brazil?)		1867	Assunguy	1874+		FO128/49; *Assunguy*
Bartholome Coyne (27)		Mary (23)		"Irish"	1867/8	Príncipe Dom Pedro	1870		FO128/94/243; FO128/94/376; FO128/95/141

British immigrants in Príncipe Dom Pedro, Cananéia and Assunguy – *continued*

NAME (age on arrival)	Occupation	Marital status	Family composition (age on arrival)	Place of origin	Arrival year	Colony	Depart year	Last known destination	Sources
John Crossley				Warwickshire?	1872	Assunguy			LUC 30/11/72; Times 24/12/72
Albert Cull (27)	labourer	single			1872/3	Curitiba [Assunguy]	1873	Rio/?	FO128/98/506
John H. Curson					1868	Assunguy	1874?		APP 0387 p. 166
George Curtis					1868	Assunguy	1870?		FO128/92/260; APP 0387 p. 163
Daniel Daly					1868	Cananéia	1870+		ABT 7/11/68; FO128/94/294
John Daughty (37)		Emma (35)	1 child Louise? (d. Rio)		1873	Cananéia	1873	England	PP 1874; Times 15/10/73; FO128/101/149; FO128/100/174
Charles Davies (34)		Lucy (28)	James (8), Emily (7), Jessie (5), Sarah (3), Lucy (2)				1873	England	Times 15/10/73; FO128/100/174
George Davies (26)		Ann (21)	Henry (6), Kate (3) (plus 1 b. Brazil)		1872	Cananéia?	1873	England	Times 15/10/73; FO128/100/169

Appendix 3

British immigrants in Príncipe Dom Pedro, Cananéia and Assunguy – *continued*

NAME (age on arrival)	Occupation	Marital status	Family composition (age on arrival)	Place of origin	Arrival year	Colony	Depart year	Last known destination	Sources
James Davies	sawyer	Bridget Garity	6 sons (including William, 9, Thomas, James and Walter), 2 daughters (some b. Brazil)	Birmingham (Bridget, b. Westmeath, Ireland)	1868	Cananéia	1898+	James returned to Birmingham in 1886; d. 1887	*ABT* 7/11/68; *Cananea*; FO128/94/294; FO128/117; FO128/138; FO128/146; FO128/150; FO128/157; FO128/240
James Davies		single		Ashelworth Lawn, Gloucestershire	1872	fell ill in Rio	1874+	Cachoeira, Bahia	FO268/28
John Frederick Davis	railway worker	Henriette d. Bariguy	1 child (d.)	Gloucestershire	1872/3	Curitiba/Assunguy?	1873	Cirencester	*WGS* 4/10/73; *Times* 15/10/73
James Dawkins		married	5 children	Leicestershire	1872	Cananéia	1873+	Mendes, RJ/England?	FO128/101/303; PP 1873
George Dickens (33)		Sarah (34)	Elizabeth (9), Caroline (8), Margaret (7), Matilda (3), Arthur (baby)		1872		1873	England	*Times* 15/10/73; FO128/100/174
Patrick Dobbins [or Dobbings]		married	5 children	"an Irishman from Pennsylvania"	1867	Assunguy	1881+		FO128/116; *LUC* 16/11/72; *Assunguy;* APP 0387 p. 116
Patrick Donohue		Jane			1868	Príncipe Dom Pedro	1868+		*ABT* 8/7/1869

British immigrants in Príncipe Dom Pedro, Cananéia and Assunguy – continued

NAME (age on arrival)	Occupation	Marital status	Family composition (age on arrival)	Place of origin	Arrival year	Colony	Depart year	Last known destination	Sources
Henry Doughty		married		Leamington, Warwickshire?	1872	Cananéia	1873+	Mendes, RJ/ England?	*LUC* 26/7/73; FO128/101/303; PP 1874
John Doughty		married		Leamington, Warwickshire?	1872	Cananéia?	1873+		*LUC* 26/7/1873
William Drinkwater		single			1873	Cananéia	1873	Rio	PP 1874; FO128/101/349
Richard East		Elizabeth	6 children	Gloucestershire?	1872/3	Assunguy	1874	England	FO128/102/294; *WGS* 19/7/73
Charles Elliot (32)		Ann (30)	Agnes (1) plus 1 d.		1872	Cananéia	1873	England	FO128/100/175; FO128/101/303; PP 1874; *Times*, 15/10/73; APP 0387 p. 172
Samuel Elliot					1868	Assunguy	1876+		FO128/109/349
Mr Ely				New York	1867	Príncipe Dom Pedro			FO128/94/376
John Emmery						Assunguy			APP 0382 p. 18
James Estcourt	"in corn trade"			Gloucester	1872	Cananéia	1873+	Mendes, RJ/ England?	FO128/101/300; PP 1874
George Estcourt					1872	Cananéia	1873+		FO128/101/303

Appendix 3

British immigrants in Príncipe Dom Pedro, Cananéia and Assunguy – *continued*

NAME (age on arrival)	Occupation	Marital status	Family composition (age on arrival)	Place of origin	Arrival year	Colony	Depart year	Last known destination	Sources
James Faggart (35)					1867/8	Príncipe Dom Pedro			FO128/94/244
John Farmer					1868	Assunguy	1874?	São Paulo	APP 0387 p. 168
William Farrell (36)		Bridget (32)	Joanna (13), Ellen (4), William (2)		1868	Príncipe Dom Pedro			FO128/94/243
William Faville (d. 1877+)		Fanny Faville (dress maker)	2 children			Assunguy	1888+		FO128109/463; FO128/111/97; FO128/155
Thomas Fell	farm labourer, lastly rail worker, Rugby	married (d.)	1 child (d.)	Napton-on-the-Hill, Warwickshire	1872	Cananéia	1873	New York – to Tioga, Pennsylvania	*LUC* 3/5/73; *LUC* 1/11/73; FO128/101/303; *Times*, 7/4/73; PP 1874
Martin Ferrick (24)		Ellen (21)			1868	Príncipe Dom Pedro	1870+		FO128/94/243
Thomas Fisher		married	4 children		1873	Cananéia	1873	Rio/?	PP 1874
Michael Fitzgerald (24)	groom	Mary (20)	John (1), Maria (b. Brazil, 1869) plus 3 b. Brazil)	"Ireland"	1868	Príncipe Dom Pedro/ Assunguy 1870?	1870 / 1888+	Assunguy	FO128/94/243; FO128/95/141; FO128/155; *Assunguy*; APP p. 232

British immigrants in Príncipe Dom Pedro, Cananéia and Assunguy – continued

NAME (age on arrival)	Occupation	Marital status	Family composition (age on arrival)	Place of origin	Arrival year	Colony	Depart year	Last known destination	Sources
P.W. Fitzgerald		married	6 children		1868	Príncipe Dom Pedro	1869	Rio/?	FO28/92/95
Martin Flaming (40)		Sarah (30)	Michael (13)		1867/8	Príncipe Dom Pedro			FO128/94/243
Thomas Follett					1872	Cananéia	1873+	Mendes, RJ/ England?	FO128/101/303; PP 1874
Peter Forbes (37)	farm labourer	Margaret (33)	7 children, ages 1 to 14 (some b. Brazil?)		1872/3	Curitiba/ Assunguy	1873	Rio/?	Assunguy; FO128/98/506
Elijah Fulcher		married	5 children		1871/2	Cananéia	1874+		Cananea; FO128/100/198
John Gibbs		married	3 children		1873	Cananéia	1874+		Cananea
Joseph Gibbs		married	1 son, 1 daughter		1872/3	Cananéia	1876+		FO128/112/13; Cananea
John [or James?] Gillingham (35)		Phoebe (36)	Thomas (15), Daniel (13), Emily (11), Bessie (7), Frederick (3)	(Piddletown, Dorset?)	1872	Cananéia	1873	England	PP 1874; Times 15/10/73; FO128/100/174
Mr Glace				Wednesbury	1868	Príncipe Dom Pedro		England	FO128/112/13

Appendix 3

British immigrants in Príncipe Dom Pedro, Cananéia and Assunguy

NAME (age on arrival)	Occupation	Marital status	Family composition (age on arrival)	Place of origin	Arrival year	Colony	Depart year	Last known destination	Sources
William Godwin (d. 1873)		Ellen (or Rose?)	6 children			Assunguy	1874	Paranaguá/ England?	FO128/49; FO128/103/38; FO128/105/88; APP 0382 p. 18
Mr Goodman		single				Assunguy	1874+		*Assunguy*
Jesse Goode (d. Cananéia)				Mollington, Oxfordshire	1872	Cananéia	d. 1872		RLSC 23/11/72; LUC 21/12/72
William Gordon	engine driver	married	3 children	Gloucester	1872/3	Assunguy	1874+		
Thomas Gould		married (met in Brazil?)	6 children (all b. Brazil?)	previously resident in New York	1867	Assunguy	1874+		FO128/49; *Assunguy*; APP 0387 p. 114
Thomas Grady		married	2 children				1874	England	FO128/102/294
Mr Graham		single				Assunguy	1874+		*Assunguy*
Edwin Grant (39)			Frederick (15), William (13), Alexander (9)				1873	England	*Times* 15/10/73; FO128/100/169
Ernest B. Grimes					1868	Assunguy	1870+		FO128/92/260; FO128/92/304; APP 0387 p. 173

British immigrants in Príncipe Dom Pedro, Cananéia and Assunguy – *continued*

NAME (age on arrival)	Occupation	Marital status	Family composition (age on arrival)	Place of origin	Arrival year	Colony	Depart year	Last known destination	Sources
John Gudge [or Grudge?]					1872	Cananéia	1873	Mendes, RJ/ England?	FO128/101/303; PP 1874
Richard Guppy		married	2 children				1874	England	FO128/102/294
William Gusle?						Assunguy	1877+		FO128/111/97
Thomas Haines		married	5 children				1874	England	FO128/102/294
John Haher	chaplain				1868	Príncipe Dom Pedro	1869		FO128/90/174
George Hallbrooke		married (d.)	2 children				1874	England	FO128/102/294
John Hardie		married			1873	Cananéia	1873	Rio/?	PP 1874
Charles Hardiman [or Hardyman?]		married	2 children		1868	Cananéia	1873	England	ABT 7/11/1868; FO128/94/294; FO128/100/198
Joseph Hardy (39)		Elizabeth (38)	Joseph (17), Charles (11), Sarah (5), Frederick (2), (plus 1 b. Brazil)		1873	Cananéia	1873	England	PP 1874; *Times* 15/10/73; FO128/100/174; FO128/101/349
James Hansford		married	3 children				1874	England	FO128/102/294

Appendix 3

British immigrants in Príncipe Dom Pedro, Cananéia and Assunguy – *continued*

NAME (age on arrival)	Occupation	Marital status	Family composition (age on arrival)	Place of origin	Arrival year	Colony	Depart year	Last known destination	Sources
Thomas Harford (36)		Mary Ann (30)	Thomas (13), Henry (5) (d. Cananéia and Rio)	Clifton (Oxfordshire?)	1872	Cananéia	1873	England	*LUC* 11/1/73; FO128/100/175; PP 1874; *Times* 15/10/73
William Harper	agricultural labourer	married (d.)	4 children (some b. Brazil?)			Assunguy	1888+		FO128/109/463; FO128/155; *Assunguy*
James Hart	agricultural labourer	married	6 children		1868	Assunguy	1888+		FO128/92/261; FO128/155; APP 0387 p. 165
Mr Harrell					1867/8	Príncipe Dom Pedro	1870	Cananéia	FO128/95/186
Mr Harrington		married	6 children			Assunguy	1874+		*Assunguy*
James?. Harrison					1868?	Assunguy	1876		FO128109/463; APP 0387 p. 185
John Haskell				(arrived in Brazil c. 1849)	1867/8	Príncipe Dom Pedro	d. 1869		FO128/92/244
Samuel Hawgood				London	1868	Assunguy	1873	England	*LUC* 16/11/72; APP 0387 p. 135
Charles Hawkins		married	3 children		1872	Cananéia	1873+	Mendes, RJ/ England	FO128/101/303; PP 1874

243

British immigrants in Príncipe Dom Pedro, Cananéia and Assunguy – *continued*

NAME (age on arrival)	Occupation	Marital status	Family composition (age on arrival)	Place of origin	Arrival year	Colony	Depart year	Last known destination	Sources
John Healey (26)		Ellen (23)	Luke (4), John (1), (plus 1 b. Brazil)		1867/8	Príncipe Dom Pedro	1870+		FO128/94/243;
James Hennessey					1868	Cananéia	1870		*ABT* 7/11/1868; FO128/94/294
Arthur Herridge					1872	Assunguy			APP p. 233
William Hickman					1872	Cananéia	1873+	Mendes, RJ/ England?	FO128/101/303; PP 1874
Elizabeth Hill		married d. Cananéia	(1 b. Brazil)		1872	Cananéia	1873	Rio/?	PP 1874
Robert Hill (29)		Hariett (30)	Hariett (7), Mary (6), Ellen (3) Edward (baby)		1872	Cananéia	1873	England	FO128/100/173; FO128/101/303; PP 1874; *Times* 15/10/1873
James Hoode		married					1874	England	FO128/102/294
Anthony Hooly (41)					1867/8	Príncipe Dom Pedro			FO128/95/244
Mr Hopkins		married (d. 1869)	2 children (d. Príncipe Dom Pedro, 1869)		1867/8	Príncipe Dom Pedro	1869+		FO128/92/244

Appendix 3

British immigrants in Príncipe Dom Pedro, Cananéia and Assunguy – *continued*

NAME (age on arrival)	Occupation	Marital status	Family composition (age on arrival)	Place of origin	Arrival year	Colony	Depart year	Last known destination	Sources
John Horan (25)		Catherine (23)	(1 b. Brazil, 1869)	"Irish"	1867/8	Príncipe Dom Pedro	1870		FO128/94/244; FO128/94/376; FO128/95/141
Alfred Houghton		Anne (36)	John (15), Anne (13), Margaret (12), Thomas (10), William (6), Bridget (4), Patrick (2)		1868	Príncipe Dom Pedro / Assunguy?	1870		FO128/94/243; APP 0387 p. 154
Henry House (29)		Elizabeth (28)	William (4), Maud (2)		1872		1873	England	*Times* 15/10/73; FO128/100/198
Richard House [or Howse?]	labourer	married	7 children	Oxfordshire	1873	Assunguy	1874+		*Assunguy*
Stephen Hubbard					1868	Cananéia	1868+		*ABT* 7/11/1868
William Hubie		married			1868	Assunguy	1874+		FO128/92/260; *Assunguy*; APP 0387 p. 157
Mr Hughs					1867/8	Príncipe Dom Pedro	1869+		FO128/92/244
James Hurst (or Hunt?)		married	2 children		1868	Assunguy	1869		FO128/92/260; FO128/92/263

British immigrants in Príncipe Dom Pedro, Cananéia and Assunguy

NAME (age on arrival)	Occupation	Marital status	Family composition (age on arrival)	Place of origin	Arrival year	Colony	Depart year	Last known destination	Sources
James Hutchins [or Hatching or Hitchens]		married (d. 1873, Bariguy)		Corfe Castle, Dorset	1873	Assunguy	1874	England	FO128/104/163; FO128/105/87; APP 0382 p. 150
Joseph Jackson		married	1 child (b. Brazil?)			Cananéia	1874+		*Cananea*
William Jessetts		single		Napton, Warwickshire	1872	Cananéia	1873+		*LUC* 8/2/1873; *LUC* 15/2/73; FO128/100/198
Edward Jew					1872	Cananéia	1873+		FO128/101/303
William A. Johnston		married	1 child		1873	Cananéia	1874+		*Cananea*
Walter Joslin					1868	Assunguy	1869		APP 0387 p. 169
Honoria Joyce		widow	4 sons, 1 daughter			Cananéia	1878+		*Cananea*; FO128/116; FO128/100/198
George Kendrick (35)	Harriet (35)		William (13), Sydney (10), Annie (8), Albert (6), Lizzie (4)				1874		FO128/104/167
Joseph Kendrick			2 children (d. Bariguy and Antonina)	Castle Norton, Worcestershire	1873	Curitiba/ Assunguy	1874	Paranaguá	FO128/104/166

Appendix 3

British immigrants in Príncipe Dom Pedro, Cananéia and Assunguy – *continued*

NAME (age on arrival)	Occupation	Marital status	Family composition (age on arrival)	Place of origin	Arrival year	Colony	Depart year	Last known destination	Sources
Peter Kenny (35)		Bridget (32)	Mary (9), Honora (7), James (3), John (1)		1868	Príncipe Dom Pedro	1870+		FO128/94/243
Cornelius Kirby (or Kerby) (28)					1867/8	Príncipe Dom Pedro	1870		FO128/94/244; FO128/95/141
James Kirby (or Kerby) (40)		Alice (36)	James (20), Catherine (17), Anna (14), Honora (9), Bridget (7), Margaret (5), Mary (3), John (1)	"Irish"	1868	Príncipe Dom Pedro	1870		FO128/94/243; FO128/94/375; FO128/95/141
Joseph Kilchin [or Kinchen?] (32)	tool maker	Mary	6 children	Sharon Vale, nr Sheffield, last of "nr Birmingham"	1872	Cananéia	1887+?	Mrs Kilchin working in Santos, 1887	FO128/112/13; *Cananea; Times* 24/12/74; FO128/150
Miles Kirby (or Kerby) (35)		Mary (28)	James (14), Cornelius (8), Honora (5), Bridget (1), Patrick (b. Brazil 1869)		1868	Príncipe Dom Pedro	1870		FO128/95/141
Anne Knaughten (34)		married d.	John (14), Anne (12), Margaret (10), Thomas (8), Bridget (6), Patrick (1)	"Irish"	1867/8	Príncipe Dom Pedro	1870		FO128/94/376; FO128/95/141

British immigrants in Príncipe Dom Pedro, Cananéia and Assunguy – *continued*

NAME (age on arrival)	Occupation	Marital status	Family composition (age on arrival)	Place of origin	Arrival year	Colony	Depart year	Last known destination	Sources
Bridget Knaughten (38)		married d.			1867/8	Príncipe Dom Pedro	1870		FO128/95/141
John Landner	carter	Jane	"children"	Ampney and Daglingworth, nr Cirencester, Gloustershire	1873	Curitiba/ Assunguy	1873+	Antonina	WGS 13/09/73; FO128/101/319
Miss Lane						Assunguy	1888+		FO128/155
Albert Lane	worked with Gloucester Wagon Co.			Gloucester	1872	Cananéia	1873+	Mendes, RJ/ England?	FO128/101/303; FO128/101/300; PP 1874
Allen Leightholders					1872	Assunguy	1873+		FO128/49
Patrick Leighton (or Leyden)	labourer	Mary	5 children (some b. Brazil?)		1869	Assunguy	1874+		*Assunguy*; APP p. 231
H. A. Levvy					1868	Cananéia	1868+		*ABT* 7/11/68
Robert Lewis (29)					1867/8	Príncipe Dom Pedro			FO128/94/244
Mr Lightless		married				Assunguy	1874+		*Assunguy*

248

Appendix 3

British immigrants in Príncipe Dom Pedro, Cananéia and Assunguy – *continued*

NAME (age on arrival)	Occupation	Marital status	Family composition (age on arrival)	Place of origin	Arrival year	Colony	Depart year	Last known destination	Sources
Patrick Lowden (31)		Maria (29)	William (4), Martin (2) (plus 1 b. Brazil)	"Irish"	1868	Príncipe Dom Pedro	1870		FO128/94/243; FO128/94/376; FO128/95/141
William Lomas		single			1872	Assunguy	1876+		FO128109/463; *Assunguy*
William Loveridge (27)	carpenter	Jane (19)	(1 child b. Brazil)	Barton St Michael, Gloucestershire	1873	Assunguy	1874	England	*Assunguy*; family records
Patrick Lyden		married	5 children			Assunguy	1874+		FO128/49; *Assunguy*
Henry Lynn d. Bariguy?		Lavinia	5 children			Curitiba/ Assunguy	1874	Paranaguá; children adopted in Curitiba	FO128/104/48; APP 0453 p. 5
Edward Lyons (52)	lamp maker	married	4 children	Birmingham	1872	Cananéia	1876+	Rio de Janeiro/ England?	*LUC* 21/12/72; *Cananea*; *Times* 24/12/72; FO128/112/21
Bridget McAndrew					1867/8	Príncipe Dom Pedro	1869		FO128/94/243
Michael McComville	soldier/ library caretaker	married	5 children	Belfast	1872?	Cananéia	1874+		PP Brazil, 1873; *Cananea*

British immigrants in Príncipe Dom Pedro, Cananéia and Assunguy – continued

NAME (age on arrival)	Occupation	Marital status	Family composition (age on arrival)	Place of origin	Arrival year	Colony	Depart year	Last known destination	Sources
Thomas McDermott					1867/8	Príncipe Dom Pedro	d. 1869		FO128/92/245
John McDonald					1868	Assunguy	1869+		FO128/92/260; APP 0387 p. 160
Patrick MacGary [or McGarry]	labourer	married	5 children (at least 1 daughter d., son "mad")	Ireland	1872	Assunguy	1876+		FO128/49; FO128109/463; *Assunguy*
Cornelius McGillcuddy		Mary			1868	Assunguy	1874+		FO128/92/260; *Assunguy*; APP 0387 p. 177
Alfred McHoughton					1868	Assunguy	1869+		FO128/92/260
William Malin (30)		Sarah (25)	William Charles (1)		1872	Cananéia	1873	England	FO128/100/175; PP 1874; *Times* 15/10/73
William Mansell	labourer	married	1 child	Gloucestershire	1873	Assunguy	1874+		FO128/49; FO128/106/339; *Assunguy*
Cornelius Marsh	labourer/ butcher	married d.	2 children (1 b. Brazil?)	Dorset	1872	Assunguy	1888+		FO128/109/463; FO128/155: *Assunguy*

Appendix 3

British immigrants in Príncipe Dom Pedro, Cananéia and Assunguy

NAME (age on arrival)	Occupation	Marital status	Family composition (age on arrival)	Place of origin	Arrival year	Colony	Depart year	Last known destination	Sources
Henry Mason	farmer (former sergent-major, British army?)	married	2 children (Archie and Amy?)	Lancashire	1868	Assunguy	1877+		FO128/49; FO128/92/260; FO128/92/304; FO128109/463; FO128/111/97; *Assunguy*; Tigar APP 0387 p. 170
Mr Mayley					1867/8	Príncipe Dom Pedro	1870	Cananéia	FO128/95/186
Thomas H. Meech						Assunguy	1873		FO128/49
William Mellish (35)		Emma (36)	Ellen (11), Eliza (8), Mary (6), Emily (4)		1872		1873	England	*Times* 15/10/73; FO128/100/174
Alfred Merry	labourer/ bricklayer	married	7 children (some b. Brazil)	Middleton Cheney, Northamptonshire	1872	Assunguy	1888+		*LUC* 4/1/73; FO128/155; *Assunguy*
James Messenger (27)		Henriette (24)	Richard (4)		1872		1873	England	*Times* 15/10/1873; FO128/100/174
Edward Michell [or Mitchell?]					1868	Cananéia	1870+		*ABT* 7/11/68; FO128/94/294
James Millard	railway carrier			Gloucester	1872	Cananéia	1873	New York	FO128/101/300; *Times* 7/4/73; PP 1874

251

British immigrants in Príncipe Dom Pedro, Cananéia and Assunguy – *continued*

NAME (age on arrival)	Occupation	Marital status	Family composition (age on arrival)	Place of origin	Arrival year	Colony	Depart year	Last known destination	Sources
Alfred Mills (20)					1872/3	Curitiba/ Assunguy	1874	England	FO128/104/167
Thomas Miller	labourer	married	1 child	Gloucestershire	1873	Assunguy	1877+		FO128/109/463; FO128/111/97; *Assunguy*
John Mitchell	agricultural labourer	married				Curitiba/ Assunguy	1874	England	FO128/102/232; FO128/103/39
John Mitchell (45)		Martha	Edward, Samuel, Eliza		1868	Cananéia	1881+		*ABT* 7/11/68; Liverpool passport; *Cananea*; FO128/94/294; FO128/116; FO128/117; *ABT* 22/1/74; *ABT* 22/1/74
Richard Monks		married	3 children		1872/3	Cananéia	1874+		*Cananea*
William Morris					1868	Assunguy			APP 0387 p. 191
William Mortimer					1869	Assunguy			APP 0387 p. 149
James Muller					1868	Cananéia	1870+		*ABT* 7/11/68; FO128/94/294
Thomas Muller					1872	Assunguy	1876+		FO128109/463

252

Appendix 3

British immigrants in Príncipe Dom Pedro, Cananéia and Assunguy – *continued*

NAME (age on arrival)	Occupation	Marital status	Family composition (age on arrival)	Place of origin	Arrival year	Colony	Depart year	Last known destination	Sources
Cornelius Murphy (38)		Mary (32)	Mary (12), Julia (7), Cornilius (2), William (baby)		1868	Príncipe Dom Pedro/ Cananéia	1869/ 1874+		FO128/94/243; *Cananea*
Patrick Murphy (48)		Bridget (32)	Margaret (14), Michael (14), Thomas (11), William (11), John (6), Kate (2), Edward (infant)	"Irish"	1867/8	Príncipe Dom Pedro	1870	Cananéia	FO128/94/243; FO128/94/375; FO128/95/141; FO128/95/186
Michael Murray		married	3 sons, 3 daughters (some b. Brazil)	Birmingham? (b. Ireland)	1868	Cananéia	1888+		FO128/94/294; FO128/ 116; FO128/117; FO128/157; *Cananea*
J. Neal		married				Assunguy	1874+		*Assunguy*; APP 0382 p.18
Samuel Newport (d. 1874?)	carpenter	Jane Newport	3 children	Piddletown, Dorset [last Bristol?]	1872	Assunguy	1876	England	FO128/105/74; FO128107/230; *Assunguy*; APP p. 234
John Nichols		married	7 children			Cananéia	1874+		*Cananea*
Richard Noonan					1868	Cananéia	1870+		*ABT* 7/11/68; FO128/94/294
Julia Norman						Assunguy			APP 0387 p. 147

253

British immigrants in Príncipe Dom Pedro, Cananéia and Assunguy – continued

NAME (age on arrival)	Occupation	Marital status	Family composition (age on arrival)	Place of origin	Arrival year	Colony	Depart year	Last known destination	Sources
John Nottley		married	2 children	Piddletown, Dorset?	1872	Cananéia	1873	Rio/	PP 1874
J. O'Donnel		married	3 children	New York [originally from Ireland?]	1867	Príncipe Dom Pedro			FO128/94/376
Francis Ole	agricultural labourer	married	4 children (at least 2 b. Brazil)			Assunguy	1888+		FO128/155; *Assunguy*
John O'Neill (29)		Anne (27)	Bernard (8), Edward (6), Mary (3)		1867/8	Príncipe Dom Pedro	1869		FO128/94/243
Austin Orgill (26)		Martha Connolly (25)	Mary (2) and Joseph (baby) (d. Rio, 1869)	Wednesbury	1868	Príncipe Dom Pedro	1868/9	Pittsburgh, Pennsylvania	family records
William O'Rorke		married	4 children (some b. Brazil?)		1867/8?	Cananéia	1874+		*Cananea*
George Osborn		married	2 children d. Cananéia	Leamington, Warwickshire	1872	Cananéia	1873	England	PP 1874; *Times* 17/10/73
O'Sullivan, *see also* 'Sullivan'									
Michael O'Sullivan	agricultural labourer	married	James and 6 other children	Ireland	1872	Assunguy	1888+	England?	FO128/112/594; FO128/155; *Assunguy*; APP 0453 p. 372

Appendix 3

British immigrants in Príncipe Dom Pedro, Cananéia and Assunguy – *continued*

NAME (age on arrival)	Occupation	Marital status	Family composition (age on arrival)	Place of origin	Arrival year	Colony	Depart year	Last known destination	Sources
Frank Over	stonemason	married d.	5 children		1875	Assunguy	1888+		FO128109/463; FO128/111/97; FO128/155
Mr Owens					1867/8	Príncipe Dom Pedro	1869+		FO128/92/244
Frederick Painter (40)		married	Mary Ann (13), Jessie (5), George (3)		1873		1873	England	*Times* 15/10/73; FO128/100/169
J. (Hammond?) Parker		single			1872	Assunguy	1874+		FO128/49; *LUC* 28/12/72; *LUC* 4/1/73; *Assunguy*
George Payne		married	children				1879+		FO128/117
Edward Pedder					1868	Cananéia	1870+		*ABT* 7/11/68; FO128/94/294
Robert Peeling (48/9)		married d. Curitiba	Samuel (18), Liddie (13), Robert (9), Fannie (4), Elsie (2); 1 other (d. Curitiba)		1872/3	Assunguy	1874	Paranaguá/ England	FO128/103/38; FO128/104/167; FO128/105/87
Joseph Perott (28)		married d.	Charles (4), John (1)		1872/3	Curitiba/ Assunguy	1874	England	FO128/104/167
John Perrigo (41)		Elizabeth (26)	Elizabeth (12), Ann (10), Eliza (6)		1872		1873	England	*Times* 15/10/73; FO128/100/173

British immigrants in Príncipe Dom Pedro, Cananéia and Assunguy

NAME (age on arrival)	Occupation	Marital status	Family composition (age on arrival)	Place of origin	Arrival year	Colony	Depart year	Last known destination	Sources
James Petrie (20)	plasterer				1872/3	Curitiba/ Assunguy	1873	Rio/?	FO128/98/506
John Philips	labourer	married	8 children (some b. Brazil)	Gloucestershire	1873	Assunguy	1888+		FO128/109/463; FO128/155; *Assunguy*
James Pike (brother of Thomas)	carpenter/ sawyer				1868	Assunguy	1888+		FO128/49; FO128/92/261; FO128/109/463; FO128/155; *Assunguy*; APP 0387 p. 174
Thomas Pike (brother of James)	carpenter	married			1868	Assunguy	1888+		FO128/49; FO128/92/261; FO128/155; *Assunguy*; APP 0387 p. 152
Sarah Pimble		married d.	no children		1868	Cananéia	1874+		*Cananea*
Mr Pinson, d. Rio 1869		married	2 children		1868	Príncipe Dom Pedro	1869		FO128/92/95
James Piper						Assunguy			APP 0387 p. 142
Thomas Pitham		married d.	2 children (d. Cananéia)	Bilton, Warwickshire	1872	Cananéia	1873	New York	FO128/101/303; *Times* 7/4/73; PP 1874
Richard Prouse		married	4 children		1872		1873	England	*Times* 15/10/1873

Appendix 3

British immigrants in Príncipe Dom Pedro, Cananéia and Assunguy – *continued*

NAME (age on arrival)	Occupation	Marital status	Family composition (age on arrival)	Place of origin	Arrival year	Colony	Depart year	Last known destination	Sources
Henry Pugh				Gloucestershire	1872	Cananéia	1873	New York England?	FO128/101/303; *Times* 7/4/1873; PP 1874
John Pugsley (36)	blacksmith/ shipwright	Elizabeth Ackland	Alfred (11/12), Fanny (9), John (8), William (6), Elizabeth (5), George (infant), Kate (?)	Barnstaple, Devon (John recruited in Cirencester, Gloucestershire)	1872/3	Assunguy	1874+	Curitiba	family records; *Assunguy*
Joseph Pullen	labourer	Emma	Alfred plus 7 children	Gloucestershire	1873	Assunguy	1878+		FO128109/463; FO128/112/607; FO128/112/638; *Assunguy*; APP 0453 p. 263
John Pulling (53)		Caroline (47)	William (14), Thomas (11), Harriett (22) and her children (?) Emily (3) and Albert (baby)		1872	Cananéia	1873	England	FO128/100/173; *Times* 15/10/73
William Quail (or Quayle?)		married	1 child			Assunguy	1874+		*Assunguy* 1875; Tigar
Alfred Rafferty (17)					1867/8	Príncipe Dom Pedro			FO128/94/244
William Rainbow		single		Napton, Warwickshire	1872	Cananéia	1873+		*LUC* 11/1/73; FO128/100/198

British immigrants in Príncipe Dom Pedro, Cananéia and Assunguy – *continued*

NAME (age on arrival)	Occupation	Marital status	Family composition (age on arrival)	Place of origin	Arrival year	Colony	Depart year	Last known destination	Sources
James (or John?) Randall (27)	agricultural labourer	married d. Assunguy	James Frederick (3) plus 1 baby	Somerset ("nr Bristol")	1872	Assunguy	1873	England	FO128/49; *Times* 15/10/73; *Times* 15/9/73; WGS 20/9/73; FO128/100/169
George Randall (or Randell) (21)					1867/8	Príncipe Dom Pedro			FO128/94/244
Hannah Randall (49 in 1878)		married d. 1878	Robert (6), Alfred (9), William (11), Alice (12), Louisa (16), Frederick (18), George (20) – ages in 1878 (some b. Brazil)			Assunguy	1879	England	FO128/116; APP 0453 p. 332
Joseph Redman	clergyman and teacher	widow	8 children (2 d. Assunguy 1883)		1877	Assunguy	1884	Desborough, Northamptonshire	FO128/112/638; FO128/116; FO128/127; FO128/129; FO128/136; FO128/138
Joseph Renaudin	nurse		3 children	London	1872/3	Assunguy	1879+		FO128/103/452; FO128/105/314; FO128/116; *Assunguy*
William (?) Roberts		single				Assunguy	1874+		*Assunguy*; APP 0382 p. 18

Appendix 3

British immigrants in Príncipe Dom Pedro, Cananéia and Assunguy – *continued*

NAME (age on arrival)	Occupation	Marital status	Family composition (age on arrival)	Place of origin	Arrival year	Colony	Depart year	Last known destination	Sources
Henry Robins	agricultural labourer	married	5 children	Gloucestershire	1873	Assunguy	1876+		FO128/109/463; *Assunguy*
William D. Robinson		single	(married 1877)		1868	Assunguy	1880	Curitiba	FO128/108/305; FO128/111/97; FO128/116; APP 0387 p. 162
George Rochett (or Rocket?)	married		3 children	(+ William Bartlett, 15 yrs)			1874	England	FO128/49; FO128/102/294
John Rouse (46)					1867/8	Príncipe Dom Pedro			FO128/94/244
George Russell (39)		Elizabeth (36)	Elizabeth Ann (4), Catherine Mary (2)		1872	Assunguy	1873	England	FO128/100/175; PP 1874; *Times* 15/10/73
J. William Ryan (40)		Mary Anne (30)	Mary Elizabeth (11), Margaret (8), James (7), Susanna (5), Martha (3), Agnes (baby)	Lincolnshire	1873	Assunguy	1874	England	FO128/103/38; FO128/104/143; FO128/104/167; FO128/105/86
Frederick Samways (31)		Ada Sarah Gifford (31)	Herbert (8), Barbara (6), George (5), Grace (3/4), Susan (2), Frederick (infant) plus six children b. Brazil	Broadmayne or West Knighton, nr Dorchester, Dorset	1872/3	Curitiba/ Assunguy?		Curitiba	*LUC* 3/5/1873; family records

259

British immigrants in Príncipe Dom Pedro, Cananéia and Assunguy – *continued*

NAME (age on arrival)	Occupation	Marital status	Family composition (age on arrival)	Place of origin	Arrival year	Colony	Depart year	Last known destination	Sources
John Sellwood [or Selwood]			2 daughters (15 and 22 in 1877)			Assunguy	1878+		FO128/112/638; *Assunguy*
James Shanahan [or Shannon?] (29)		Catherine (28)	Margaret (9), William (5), Mary Anne (4), James (3), Catherine (1). (Others b. Brazil.)	"Irish" – (Limerick?) Wednesbury?	1868	Príncipe Dom Pedro/ Assunguy?	1870+	Assunguy? 1870 / 1888+	FO128/94/243; FO128/94/376; FO128/95/141; FO128/155; *Assunguy*
John Shanahan (41)		Mary (39)	Margaret (21), Mary (17), Edward (15), William (12), John (10), Patrick (6), Anne (4), Cornelius (1)	"Irish" – Wednesbury?	1868	Príncipe Dom Pedro	1870	Assunguy? (1888+)	FO128/94/243; FO128/94/376; FO128/95/141; FO128/155
James Shannon					1867/8	Príncipe Dom Pedro?	1873+	Assunguy	FO128/49
Mr Sherman (31)		Anne (30)	Harold (4), Thomas (1). Florence (b. Brazil).		1867/8	Príncipe Dom Pedro	1870+		FO128/94/244
Thomas Sheasby		single		Napton, Warwickshire	1872		1873	Napton, Warwickshire	PP 1874; *RLSC* 11/1/73; *LUC* 11/1/73; FO128/101/349
Catherine Siles		married d. 1877	1 infant		1873		1877	Alexandra, Paraná/ Rio	FO128/115

Appendix 3

British immigrants in Príncipe Dom Pedro, Cananéia and Assunguy – *continued*

NAME (age on arrival)	Occupation	Marital status	Family composition (age on arrival)	Place of origin	Arrival year	Colony	Depart year	Last known destination	Sources
John Size (23)		Mary (23)	Mary (1), (plus 6 b. Brazil)	"Irish"	1867/8	Príncipe Dom Pedro/ Assunguy	1870 / 1888+		FO128/94/243; FO128/94/376; FO128/95/141; FO128/109/463; FO128/155 *Assunguy*; APP p. 230
Harry Smith (20)	labourer				1872/3	Curitiba/ Assunguy	1873		FO128/98/506
Percy Smith					1868	Assunguy			APP 0387 p. 182
Richard Smith	agricultural labourer	married	6 children		1868	Assunguy	1888+		FO128/92/261; FO128/109/463; FO128/155; APP 0387 p. 176
Robinson Smith		married	4 children				1874	England	FO128/102/294
Thomas Smith (24)		Sophia (21)	1 child		1872	Cananéia	1873	England	FO128/100/175; FO128/101/303; *Times* 15/10/73
George Squires (d. 1873+)	agricultural labourer	Martha	3 children			Assunguy	1888+		FO128/49; FO128/155; *Assunguy*

British immigrants in Príncipe Dom Pedro, Cananéia and Assunguy

NAME (age on arrival)	Occupation	Marital status	Family composition (age on arrival)	Place of origin	Arrival year	Colony	Depart year	Last known destination	Sources
William Stanton	coachman/ butler	Mary	1 baby d.	Great Bourton, Oxfordshire	1872	Cananéia	1872	England	FO128/101/321; RLSC 11/1/73; LUC 21/12/72; LUC 4/1/73
Janet Stone (26)		widow	David (3)			Cananéia (?)	1873	England	FO128/100/169
William Stothers					1867/8	Príncipe Dom Pedro	1869+		FO128/92/244
Mr Stout						Assunguy	1874+		*Assunguy*
George Stowe	labourer	married	"children" – (2 of whom d.)	Middleton Cheney, Northamptonshire	1872	Cananéia	1873+		LUC 4/1/1873; FO128/100/198
Patrick Sullivan						Assunguy	1873+		FO128/49
Sullivan, *see also* 'O'Sullivan'									
Patrick Sydon	agricultural labourer	married	3 children			Assunguy	1888+		FO128/155

Appendix 3

British immigrants in Príncipe Dom Pedro, Cananéia and Assunguy – *continued*

NAME (age on arrival)	Occupation	Marital status	Family composition (age on arrival)	Place of origin	Arrival year	Colony	Depart year	Last known destination	Sources
Charles Tamplin (47) d. Assunguy 187X	merchant	Caroline Tamplin (48)	Albert Cowper (26), Elizabeth Kathleen (13), Edward Maxwell (12), Charles (7), Mildred (5), Frederick (3) (plus 3 b. Brazil – 2 d. Assunguy)	London	1868	Assunguy	1881	Caroline Tamplin to Curitiba. All surviving children left Assunguy.	FO128/108/286; FO128/116; *Assunguy*; APP 0453 p. 337; family records
Stephen Terry [or Tenny?]					1868	Cananéia	1870+		FO128/94/294
Ubald Tetu	medical doctor	Sophie	5 children	Québec, Canada c. 1870		Assunguy	1874+		*Assunguy*; APP 0387 p. 202; family records
William Thorne		married	1 child			Assunguy?	1874	England	FO128/102/294
Mathew Thorpe (22)		Sarah	Ida (5), Sidney (3)		1872		1873	England	*Times* 15/10/73; FO128/100/173
Frederick Tertius Tigar (25)	engineer	Kathleen Tamplin (daughter of Charles Tamplin)	(3 children b. Assunguy)	Beverley, Yorkshire	1868	Assunguy	1880	London/ Manitoba	FO128/49; FO128/105/70; FO128/116; Tigar FO128/121; *Assunguy*
William Tilling (24)		Francis (24)	Alice (baby), Elizabeth (b. Brazil)		1867/8	Príncipe Dom Pedro	1870+		FO128/94/243

British immigrants in Príncipe Dom Pedro, Cananéia and Assunguy – continued

NAME (age on arrival)	Occupation	Marital status	Family composition (age on arrival)	Place of origin	Arrival year	Colony	Depart year	Last known destination	Sources
John Tinson (37)		Ann (35)	John (11), Sarah Ann (8), Samuel (5) (plus 2 d. Cananéia)		1872	Cananéia	1873	England	FO128/100/175; PP 1874; *Times* 15/10/1873
Charles Tinson					1868	Assunguy	1874?+		FO128/92/260; APP 0387 p. 152
Henry Tinson					1868	Assunguy	1869+		FO128/92/260
Edward Tomlin (29)		Emily (27)	Frederick (6), Arthur (4), Walter (1), plus 2(?) b. Brazil		1872		1873	England	*Times*, 15/10/73; FO128/100/174
Mr Travelle		married	3 children			Assunguy	1874+		*Assunguy*
George Travers		married	3 children		1872	Cananéia	1873	England	FO128/101/303; PP 1874; *Times* 15/10/73
John Travers (26)		Anne (25)	William (4), George (2), Christine (baby)		1872	Cananéia	1873	England	FO128/101/303; FO128/101/173
James Tripp	labourer	married	3 children	Berkshire	1872/3	Assunguy	1879?		APP 0453 p. 1
Joseph Tuplitt					1868	Assunguy			APP 0387 p. 188
John Turner				New York	1867	Cananéia	1869	São Paulo	FO128/92/260; FO128/100/165
Thomas Tustin			2 children				1874	England	FO128/102/294

264

British immigrants in Príncipe Dom Pedro, Cananéia and Assunguy – continued

NAME (age on arrival)	Occupation	Marital status	Family composition (age on arrival)	Place of origin	Arrival year	Colony	Depart year	Last known destination	Sources
John Varley (36)		Grace (32)	Anne (13), Thomas (11), Grace (5), Catherine (1), (plus 1 b. Brazil, 1869)		1867/8	Príncipe Dom Pedro	1870		FO128/94/243; FO128/95/141
James Wadham (53)		Sarah (47)	Robert (18), Charles (15), George (12), Louise (8), Edward (6)		1872	Cananéia	1873	England	FO128/100/173; FO128/101/303; PP 1874; *The Times* 15/10/73
John Walker (d.)	agricultural labourer	Elizabeth	6 children		1872?	Assunguy	1888+		FO128/155; *Assunguy*; APP p. 235
James Walker		married d. Bariguy	1 baby			Curitiba/ Assunguy	1873		
Reuben Walker	timber yard worker			Gloucester	1872	Cananéia	1873	New York	FO128/101/300; *Times* 7/4/73; PP 1874
James Wallington					1868	Assunguy	1870+		FO128/92/260; APP 0387 p. 155
Levy (Lewis) Walsh (or Welsh?), d. Rio	agricultural labourer	married d. Brazil	2 children		Curitiba/	Assunguy	1875	Rio	FO128/103/39; FO128/104/161; FO128/105/85; FO128/106/382

British immigrants in Príncipe Dom Pedro, Cananéia and Assunguy – *continued*

NAME (age on arrival)	Occupation	Marital status	Family composition (age on arrival)	Place of origin	Arrival year	Colony	Depart year	Last known destination	Sources
Thomas Walsh [or Welsh?] (29)	labourer	Bridget (29)	Bartholomu (2), Margaret (1) (plus 3 b. Brazil)	"Ireland"	1867/8	Príncipe Dom Pedro / Assunguy 1870	1888+		FO128/94/243; FO128/106/339; FO128109/463; FO128/155; *Assunguy*; APP p. 229
John Warley (37)		Grace (33)	Anne Mary (14), Thomas (11), Grace (6), Catherine (2)		1867/8	Príncipe Dom Pedro	1870+		FO128/94/243
James Watts		married	3 children			Cananéia	1874	England	FO128/102/188; FO128/100/198
William Watts		married	2 children	Middleton Cheney, Northamptonshire	1872	Cananéia	1874+		*LUC* 4/1/73; FO128/112/13; *Cananea*
George Web (25)		Fanny (22)	1 child (d. Cananéia) (plus 1 b. Brazil)		1873	Cananéia	1873	England	PP 1874; *Times*, 15/10/73; FO128/100/169
James Webb		married	1 child (d. Cananéia)		1873	Cananéia	1873		PP 1874
Sidney Webb	labourer	single		Somerset	1872	Assunguy	1874+		*Assunguy*
Thomas Webb	agricultural labourer	married	7 children (some b. Brazil)	Wiltshire	1873	Assunguy	1888+		FO128/111/97; FO128/155; *Assunguy*
William Wells (32)		Mary (35)	Mary Ann (9)		1873	Cananéia (?)	1873	England	*Times* 15/10/73; FO128/100/169

Appendix 3

British immigrants in Príncipe Dom Pedro, Cananéia and Assunguy

NAME (age on arrival)	Occupation	Marital status	Family composition (age on arrival)	Place of origin	Arrival year	Colony	Depart year	Last known destination	Sources
John Welsh	labourer	married	4 children		1873	Assunguy	1874+		*Assunguy*
William West					1868	Assunguy			APP 0387 p. 191
William Elton Westley	agricultural labourer / carpenter	married	Alfred plus 3? other children	Surrey	1865? 1869?	Assunguy	1888+?		FO128/155; APP 0453 p. 197; APP 0387 pp. 136, 227; *Assunguy*
Mr Weston		married	"children"				1874	England	FO128/102/232
Frank White (49/50?)		married d. Curitiba	Maria (17), Annie (12), James (11), Millie (8), Delilah (b. Brazil?)		1872/3	Assunguy	1874	Paranaaguá/ England	FO128/103/38; FO128/104/167; FO128/105/86
Henry White (55)		Sarah (55)	George (35)		1872	Cananéia	1873	England	FO128/100/173; *Times* 15/10/73
Charles Wickens	agricultural labourer	married	1 child (4 others d.)			Curitiba/ Assunguy	1874	Paranaguá	FO128/103/39; FO128/105/87
Thomas Wiltshire				Gloucester					
William Wirbi	agricultural labourer	married				Assunguy	1888+		FO128/155

British immigrants in Príncipe Dom Pedro, Cananéia and Assunguy – continued

NAME (age on arrival)	Occupation	Marital status	Family composition (age on arrival)	Place of origin	Arrival year	Colony	Depart year	Last known destination	Sources
William Witherington (d. 1873)		Mary Anne (33)	John (16), William, (11), Charles (9), plus 1 b. Brazil (plus 3 left in Curitiba and São José)	Poole, Dorset	1873	Assunguy	1874	Poole, Dorset	FO128/103/38; FO128/104/143; FO128/104/164; FO128/104/167; FO128/105/384
James Wood						Assunguy			APP 382 p. 18
Arthur Woodrow				Dorset	1872	Cananéia	1873	England	Times 17/10/73
George Wright		married	1 child				1874	England	FO128/102/294
Alfred Young				London		Cananéia			
Charles Henry Young	bricklayer	Susan Stammers (d. 1874)	4/5 adult children (including Edward, Ernest, Henry, Alfred, see above/below)	London	1872	Cananéia	1874+	Iguape. SP?	LUC 17/7/72; LUC 19/10/72; Cananea
Edward Young					1872	Cananéia	1873+	Iguape, SP?	Times 12/4/73
Ernest William Young (21/22)		married		London	1872	Cananéia	1874+	Iguape, SP (d. 1914)	Times 14/4/72; Cananea
Henry Young						Assunguy			APP 0382 p. 18
Henry George Young (27)		married		London	1868?	Cananéia	1874+		ABT 7/11/68; Cananea
Thomas Young		married	3 children		1868?	Cananéia	1874+		Cananea

Notes

Prologue

1 Thomas Hardy, *Tess of the D'Urbervilles* (London: Macmillan, 1963; first published 1891), p. 297.
2 *Ibid.*, p. 332.
3 *Ibid.*, p. 333.
4 *Ibid.*, p. 434.
5 *Ibid.*, p. 433.
6 *Ibid.*, p. 470.
7 Gilberto Freyre, *Ingleses* (Rio de Janeiro: José Olympio, 1942), pp. 60–5. Before the emergence of the first Republic in 1889, Brazil was divided into provinces rather than states, as later. The head of a provincial government was a President, appointed by the central government in the Brazilian capital, Rio de Janeiro.

Introduction

1 William Scully, *Brazil; Its Provinces and Chief Cities; The Manners and Customs of the People; Agricultural, Commercial and Other Statistics, Taken From the Latest Official Documents; with a Variety of Useful and Entertaining Knowledge, Both for the Merchant and Emigrant* (London: Murray & Co., 1866), p. xii.
2 For a survey of immigration schemes in Latin America attracting British or Irish emigrants, see D.C.M. Platt 'British agricultural colonization in Latin America, Part 1', *Inter-American Economic Affairs* 18/3 (1964), 3–38 and D.C.M. Platt, 'British agricultural colonization in Latin America, Part 2' *Inter-American Economic Affairs* 19/1 (1965), 23–42.
3 See Charlotte Erickson, *Invisible Immigrants: The Adaptation of English and Scottish Immigrants in Nineteenth-Century America* (Ithaca, NY: Cornell University Press, 1990); Maldwyn A. Jones, 'The background to emigration from Great Britain in the nineteenth century', *Perspectives in American History* 7 (1973), 3–92; Philip A.M. Taylor, *The Distant Magnet: European Emigration to the U.S.A.* (New York: Harper & Row, 1971), pp. 66–90.
4 Dudley Baines, *Migration in a Mature Economy: Emigration and Internal Migration in England and Wales, 1861–1900* (Cambridge: Cambridge University Press, 1985), pp. 279 and 282.
5 Some would-be emigrants did not even make it beyond the expected port of embarkation before they turned to return home. Such were the experiences of groups of emigrants from Luxembourg in 1828 and 1846–52. Hoping to start new lives in Brazil, many of the would-be emigrants were robbed in Bremen and Dunkirk and

returned home penniless and humiliated. See Claude Wey, 'Luxembourgers in Latin America and the permanent threat of failure: "return migration" in the social context of a European micro-society', *AEMI Journal* 1 (2003), 4–5.

6 Philip A.M. Taylor, *'The Distant Magnet* after twenty-five years: an essay in self-criticism', *Journal of American Studies* 31/2 (1997), 302. On the positioning of the question of re-emigration within the wider international migration research agenda, see the now much reproduced contribution by Frank Thistlethwaite, 'Migration from Europe overseas in the nineteenth and twentieth centuries', in *Rapports V Histoire Contemporaine, XIe Congres International des Sciences Historiques*, Stockholm (Göteborg: Almqvist & Wiksell, 1960), p. XX.

7 Within the context of Latin America, several studies have touched on third country migration. On English and Irish labourers leaving behind economic depression in New South Wales (Australia) for Chile, see G.J. Abbott, 'The emigration to Valparaiso in 1843', *Labour History* (Canberra) 19 (1970), 1–16. The utopian socialist migration from Australia to Paraguay in the 1890s included many people (most notably the movement's leader, William Lane) who had previously migrated from England to Australia; see Gavin Souter, *A Peculiar People: The Australians in Paraguay* (Sydney: Sydney University Press, 1981). On Welsh immigrants who had failed to secure farmland in Patagonia and opted for resettlement on the Canadian prairies, see Lewis H. Thomas, 'Welsh settlement in Saskatchewan', *Western Historical Quarterly* 4/4 (1973), 435–49. Waves of Dutch-German Mennonites from Russia and Canada sought isolation and sanctuary in Latin American during the course of the twentieth century, often involving remigration; Larry Towell's *The Mennonites: A Biographical Sketch* (London: Phaidon, 2000) is a remarkable photographic study linking Mexico with Canada. On Korean migrant networks connecting Argentina, Brazil and Paraguay with the United States, see Kyeyoung Park, '"I am floating in the air": creation of a Korean transnational space among Korean-Latino American re-emigrants', *Positions: East Asia Cultures Critique* 7/3 (1999), 667–95. European Jews seeking sanctuary in the 1930s found it increasingly difficult to find places of refuge, but some Latin American countries continued to issue settlement visas. Arriving in countries not traditionally associated with Jewish immigration, many of these immigrants hoped that they would soon be able to move to the United States, or elsewhere, where they could benefit from pre-existing networks of friends or family as well as what they perceived as generally greater opportunities; see Leo Spitzer, *Hotel Bolivia: The Culture of Memory in a Refuge from Nazism* (New York: Hill and Wang, 1998).

8 The number of steerage passengers recorded as having proceeded to Brazil from British ports between 9 April 1868 and December 1868 was only 519, and between January and November 1869 the figure was only 300. George Lennon Hunt to George Buckley Mathew, 15 June 1869, NA/PRO (National Archive/Public Record Office) FO128/92/93; Murdoch (Emigration Board, London) to Sir F. Sandford (Foreign Office), 17 December 1869, NA/PRO FO129/93/3.

1 Agricultural Colonization in Brazil, 1808–67

1 Leslie Bethell, 'The independence of Brazil', in Leslie Bethell, ed., *Brazil: Empire and Republic, 1822–1930* (Cambridge: Cambridge University Press, 1989), pp. 18–20, 38. For a general examination of Britain's relationship with Latin America, see Leslie Bethell, 'Britain and Latin America in historical perspective', in Victor Bulmer-Thomas, ed., *Britain and Latin America: A Changing Relationship* (Cambridge: Cambridge University Press, 1989), pp. 1–24.

2 See Richard Graham, *Britain and the Onset of Modernization in Brazil, 1850–1914*

Notes

(Cambridge: Cambridge University Press, 1968); Rory Miller, *Britain and Latin America in the Nineteenth and Twentieth Centuries* (London: Longman, 1993).

3 There are few studies examining the social history of British communities in Brazil. Gilberto Freyre's *Ingleses no Brasil: aspectos da influencia Britânica sobre a vida, a paisagem e a cultura do Brasil* (Rio de Janeiro: José Olympio, 1948) remains important for the first half of the nineteenth century. On the recruitment and importance of British sailors, see Brian Vale, 'British sailors and the Brazilian Navy, 1822–1850', *The Mariner's Mirror* 80/3 (1994), 312–25. For a detailed study of an individual merchant community and its interaction with Brazilian society, see Louise Guenther, *British Merchants in Nineteenth-Century Brazil: Business, Culture, and Identity in Bahia, 1808–50* (Oxford: Centre for Brazilian Studies, University of Oxford, 2004). On an important mining community in Minas Gerais, see Marshall C. Eakin, *British Enterprise in Brazil: The St. John d'el Rey Mining Company and the Morro Velho Gold Mine, 1830–1960* (Durham, NC: Duke University Press, 1989). Useful detail on the makeup of British communities is found in José Antônio Gonsalves de Mello, *Ingleses em Pernambuco: História do Cemitério Britânico do Recife e da participação de ingleses e outros estrangeiros na vida e na cultura de Pernambuco, no período de 1813 a 1909* (Recife: Instituto Arqueológico, Histórico e Geográfico Pernambucano, 1972) and Francisco Riopardense de Macedo, *Ingleses no Rio Grande do Sul* (Porto Alegre: Edições a Nação, 1975).

4 Quoted in George P. Browne, 'Government immigration policy in imperial Brazil, 1822–1870', unpublished PhD thesis, The Catholic University of America, 1972, p. 323.

5 Jeffrey Lesser 'Neither slave nor free, neither black nor white: the Chinese in early nineteenth century Brazil', *Estudios Interdisciplinarios de América Latina y el Caribe* 5/2 (1994), 23–34.

6 Martin Nicoulin *A Gênese de Nova Friburgo: emigração e colonização suíça no Brasil, 1817–1827* (Rio de Janeiro: Fundação Biblioteca Nacional, 1996).

7 On immigration to Espírito Santo, see Gilda Rocha, *Imigração estrangeira no Espírito Santo, 1847–1896* (Vitória: Privately published, 2000).

8 For a discussion of racial determinist thought in Europe and Brazil, see Víanna Moog, *Bandeirantes and Pioneers* (New York: George Braziller, 1964); Thomas E. Skidmore, *Black into White: Race and Nationality in Brazilian Thought* (Durham, NC: Duke University Press, 1993); Robert Conrad, 'The planter class and the debate over Chinese immigration to Brazil, 1850–1893', International Migration Review 9/1 (1975), 41–55; Jeffrey Lesser *Negotiating National Identity: Immigrants, Minorities, and the Struggle for Ethnicity in Brazil* (Durham, NC: Duke University Press, 1999), pp. 13–39.

9 Moog, *Bandeirantes and Pioneers*, p. 12.

10 Eduardo Silva, *Barões e escravidão* (Rio de Janeiro: Nova Fronteira, 1984), pp. 217–20, citing Luiz Peixoto de Lacerda Werneck, *Idéias sobre colonização precedidas de uma succinta exposição dos princípios geraes que regem a população* (Rio de Janeiro: Eduardo e Henrique Laemmert, 1855), a collection of essays originally published in the Rio de Janeiro newspaper *Jornal do Comércio*.

11 *Ibid.* See also Conrad, 'The planter class and the debate over Chinese immigration to Brazil, 1850–1893', 41–55.

12 See Juvêncio Saldanha Lemos, *Os mercenários do Imperador: a primeira corrente imigratória alemã no Brasil* (Porto Alegre: Palmarinca, 1993), pp. 282–308; Neill Macaulay, *Dom Pedro: The Struggle for Liberty in Brazil and Portugal, 1798–1834* (Durham, NC: Duke University Press, 1986), pp. 207–11; Platt, 'British colonization in Latin America, Part 1', 21–2.

13 John Pascoe Grenfell (Liverpool) to Edward Pascoe Grenfell, 6 February 1868,

Grenfell Papers, Corrêa do Lago archive.
14 Michael G. Mulhall, *Rio Grande do Sul and its German Colonies* (London: Longman, Green and Co., 1873), pp. 79–82; Marian Mulhall, *Between the Amazon and Andes, or Ten Years of a Lady's Travels in the Pampas, Gran Chaco, Paraguay, and Matto Grosso* (London: Edward Stanford, 1881), pp. 62–3; Hilary Murphy 'When Wexford farmers emigrated to Brazil', *Journal of the Irish Family History Society* 2 (1986), 41–3; R. Bryn Williams, *Y Wladfa* (Cardiff: University of Wales Press, 1962), pp. 7–14; Glyn Williams, 'La imagen sobre América Latina en Gales durante los siglos XIX y XX', *Estudios Latinoamericanos* (Warsaw) 6/2 (1980) 370–7.
15 Rio Grande do Sul, *Relatório do Vice-Presidente da Provincia de S. Pedro do Rio Grande do Sul, Luiz Aloes Leite de Oliveira Bello, na Abertura da Assembléa Legislativa Provincial em 01 de Outubro de 1852* (Porto Alegre, 1852), pp. 14–15; Mulhall, *Rio Grande do Sul and its German Colonies*, pp. 51–2, 170–1; Mulhall, *Between the Amazon and Andes*, pp. 52–3; Murphy, 'When Wexford farmers emigrated to Brazil', 41–3.
16 Silva, *Barões e escravidão*, pp. 217–20.
17 Emilia Viotti da Costa, *The Brazilian Empire: Myths and Histories* (Chicago: University of Chicago Press, 1985), pp. 94–124; Hermann Kellenbenz and Jürgen Schneider, 'A imagem do Brasil na Alemanha do século XIX: impressões e estereótipos da independencia ao fim do monarquia', *Estudios Latinoamericanos* (Warsaw) 6/2 (1980), 78–81; Mack Walker, *Germany and the Emigration, 1816–1885* (Cambridge, MA: Harvard University Press, 1964), pp. 96–9, 178; Browne, 'Government immigration policy in imperial Brazil, 1822–1870', pp. 143–9.
18 'Relatório da Agência Official de Colonisação', in Brazil, *Relatório do Ministério e Secretario de Estado dos Negocios da Agricultura, Commercio e Obras Publicas* (Rio de Janeiro, 1872), Annexo D, p. 8.
19 Although migration from northern Iceland to Brazil is briefly referred to in Wilhelm Kristjanson, *The Icelandic people in Manitoba* (Winnipeg: Wallingford Press, 1965), pp. 10–1. For a detailed account of this episode (for those who are fortunate enough to be able to read Icelandic) see Thorsteinn Thorsteinsson, *Aefinyrith. Fra Islandi til Brasiliu. Frystu folkflutningar fra Northurlandi [The Adventure. From Iceland to Brazil]* (Reykjavik: Sigurgeir Frithriksson, 1937–8).
20 Brazil, *Relatório do Ministério e Secretario de Estado dos Negocios da Agricultura, Commercio e Obras Publicas* (Rio de Janeiro, 1869); Augusto de Carvalho, *O Brazil: colonisação e emigração* (Oporto: Imprensa Portugueza, 1876), p. 497; Rosana Barbosa Nunes, 'Portuguese migration to Rio de Janeiro, 1822–1850', *The Americas* 57/1 (2000), 37–61.
21 Carvalho, *O Brazil*, p. 167; Daniel P. Kidder and James C. Fletcher, *Brazil and the Brazilians* (Boston, MA: Little, Brown and Co., 1866), p. 595; William Clark Griggs, *The Elusive Eden* Frank MacMullan's Confederate Colony in Brazil (Austin: University of Texas Preso, 1987), pp. 84–5, 188 n.18; Browne, 'Government immigration policy in Imperial Brazil, 1822–1870', pp. 197–8.
22 Griggs, *The Elusive Eden* (1987), pp. 84–5, 188 n.18; *The Anglo-Brazilian Times*, 8 May 1868; William Hadfield, *Brazil and the River Plate, 1870–76* (Sutton, Surrey: W.R. Church, 1877), pp. 71–2.
23 Laura Jarnagin, 'Relocating family and capital within the nineteenth–century Atlantic World economy: the Brazilian connection', in Cyrus B. Dawsey and James M. Dawsey, eds., *The Confederados: Old South Immigrants in Brazil* (Tuscaloosa: University Press of Alabama, 1995), pp. 76–83.
24 Little is known about William Scully's Irish origins, and nothing about how he came to establish his newspaper in Brazil. This is especially surprising considering that Scully edited *The Anglo-Brazilian Times* for almost all of its nineteen years of existence, the paper appearing on a fortnightly basis from 17 February 1865 until 24

September 1884, ceasing publication shortly before the newspaper's founder left Brazil for France. However, it is thought that William Scully was born in Buolick, County Tipperary, in 1821 into a family of minor landlords. Having entered hard times with Famine, Scully apparently arrived in Brazil in the 1850s or early 1860s, initially making a living as a teacher of calligraphy. Although a Catholic by birth and upbringing, Scully married in Rio into an English Protestant family. Scully died on 14 February 1885 in Pau, France, a resort in the foothills of the Pyrenees popular with wealthy English and Americans, in particular those with disorders of the lungs. *The Times*, 26 February 1885 and personal communications from Anthony McCan 17 November 2004.

25 *The Anglo-Brazilian Times*, 23 October 1868.
26 *The Anglo-Brazilian Times*, 9 October 1866.
27 *The Anglo-Brazilian Times*, 23 October 1868.
28 Hunt (Rio de Janeiro) to Mathew, 26 January 1874, NA/PRO FO128/104/68. For a discussion of *The Anglo-Brazilian Times*, see Miguel Alexandre de Araujo Neto, 'British journalism in late 19th-century Brazil: the case of the Anglo-Brazilian Times (1865–1884)', unpublished MA thesis, Institute of Latin American Studies, University of London, 1992 and also Miguel Alexandre de Araujo Neto, 'Imagery and arguments pertaining to the issue of free immigration in the Anglo-Irish press in Rio de Janeiro', *ABEI Journal: The Brazilian Journal of Irish Studies* (São Paulo) 5 (2003), 111–27.
29 *The Anglo-Brazilian Times*, 8 June 1865.
30 *The Anglo-Brazilian Times*, 23 October 1868.
31 *The Anglo-Brazilian Times*, 9 October 1865.
32 William Scully, *Brazil*, p. xii.
33 *The Anglo-Brazilian Times*, 24 May 1865. During the first half of the nineteenth century Portuguese immigrants tended to settle permanently in Brazil, but, with the introduction of steamships, levels of return migration rose. It is certainly true, however, that the immigrants mainly settled in towns and cities, establishing businesses and working as skilled artisans. See Nunes, 'Portuguese migration to Rio de Janeiro, 1822–1850', 56–9 and Maria Beatriz Rocha-Trindade, 'Portuguese migration to Brazil in the nineteenth and twentieth centuries: an example of international cultural exchanges', in David Higgs, ed., *Portuguese Migration in Global Perspective* (Toronto: Multicultural History Society of Ontario, 1990), pp. 29–41.
34 *The Anglo-Brazilian Times*, 24 May 1865. French immigrants in Brazil await study. In Rio de Janeiro, however, they were often involved in commerce and, as elsewhere in the country, French hoteliers were fairly common. Chapter 7 of this study includes a brief discussion of the situation of French agricultural immigrants in Paraná.
35 *The Anglo-Brazilian Times*, 24 May 1865 (emphasis in the original).
36 *Ibid.*
37 *The Anglo-Brazilian Times*, 23 January 1867. Many Germans had their own views on racial determinism, often maintaining that 'Germanness' could most easily be maintained in what they regarded as a lesser civilization (such as Brazil's), rather than in a higher civilization (such as the United States'); see Walker, *Germany and the Emigration*, pp. 96–9.
38 *The Anglo-Brazilian Times*, 23 January 1867.
39 *The Anglo-Brazilian Times*, 9 October 1866.
40 *Ibid.*
41 Scully, *Brazil*; James Hurst (Rio de Janeiro) to Mathew (Rio de Janeiro), 6 February 1869, NA/PRO FO128/92/263.
42 Charles Dunlop, *Brazil as a Field for Emigration: its Geography, Climate, Agricultural Capabilities, and the Facilities Afforded for Permanent Settlement* (London:

Bates, Hendy & Co., 1866), pp. 8–9, 11, 20–2.
43 *Ibid.*, pp. 20–22;
44 Scully, *Brazil*, p. 371.
45 Daniel P. Kidder and James C. Fletcher, *Brazil and the Brazilians: Portrayed in Historical and Descriptive Sketches* (Philadelphia, PA: Childs & Peterson and London: Bates, Hendy & Co., 1857), pp. 267–8.
46 Charles Blachford Mansfield, *Paraguay, Brazil, and the Plate* (Cambridge: Macmillan & Co., 1856), p. 26.
47 Henry Thomas Buckle, *History of Civilization in England*, vol. 1, (London: John W. Parker, 1857), p. 94.
48 *Ibid.*, p. 95.
49 Skidmore, *Black into White*, p. 29.
50 *The Universal News*, 15 February 1868.

2 Brazil and the Irish Diaspora

1 For a detailed account of the famine, including an excellent bibliography concentrating on recent scholarly output but also listing key contemporary sources, see James S. Donnelly, Jr, *The Great Irish Potato Famine* (Stroud: Sutton Publishing, 2001).
2 Kerby A. Miller, *Emigrants and Exiles: Ireland and the Irish Exodus to North America* (New York: Oxford University Press, 1985), p. 103.
3 *The Irish Catholic Chronicle*, 19 October 1867.
4 For an account of one Irish group settlement experience during the 1880s, including a brief survey of earlier colonization schemes in the United States, see Gerard Moran, '"In search of the promised land": the Connemara colonization scheme to Minnesota, 1880', *Éire-Ireland* 31/3–4 (1996), 130–49.
5 *The Universal News*, 30 November 1861.
6 *The Universal News*, 30 November 1861. On the legend of the Milesians, see John Carey, 'Did the Irish come from Spain? The legend of the Milesians', History Ireland 9/3 (2001), 8–11 and Peter O'Neill, *Links Between Brazil & Ireland*: 1998/99 Survey (Rio de Janeiro: Peter O'Neill, 1999), p. 100.
7 *The Cork Examiner*, 9 December 1869; Playfair (British Consul), Algiers to Foreign Office, 22 November 1869, NA/PRO HO45/8315; Declaration by Patrick Hayes and Michael Littleton to the British Consul, Marseilles, December 1869, NA/PRO HO45/8315; Resident Magistrate, Cork to Dublin Castle, 15 December 1869, NA/PRO HO45/8315; Playfair (British Consul), Algiers, 20 December 1869, NA/PRO HO45/8315; British Consul, Marseilles, 7 February 1870, NA/PRO HO45/8315; British Consul, Paris, 9 February 1870, NA/PRO HO45/8315. See also Joëlle Annie Redouane, 'The Irish in Algeria, 1830–1930' *Éire-Ireland* 21/4 (1986), 3–10.
8 The phenomenon of the 'Yankee Irish', as Irish who migrated from the United States to Latin America were sometimes termed, has barely been examined. Jonathan Machado-Curry, however, refers to this movement in 'Indispensable aliens: the influence of engineering migrants in mid-nineteenth century Cuba', unpublished PhD thesis, London Metropolitan University, 2003.
9 See Graham Davis, *Land! Irish Pioneers in Mexican and Revolutionary Texas* (College Station: Texas A & M University Press, 2002).
10 On the background to Argentina's distinctive Irish settlement pattern, including Father Fahy's central role, see Patrick McKenna, 'Irish society in nineteenth century Argentina', in Oliver Marshall, ed., *English-Speaking Communities in Latin America* (London: Macmillan, 2000), pp. 81–103. For a demographic and economic history of the Irish in nineteenth-century Argentina, see Juan Carlos Korol and Hilda Sabato,

Cómo fue la inmigración irlandesa en Argentina (Buenos Aires: Plus Ultra, 1981). Regarding the importance (or otherwise) of 'Irishness' amongst Irish immigrants in Argentina, see Edmundo Murray, 'How the Irish became gauchos ingleses: diasporic models in Irish-Argentine literature', unpublished Mémoire de Licence, l'Université de Geneve 2003 and also Edmundo Murray, *Devenir irlandés: narrativas íntimas de la emigración irlandesa a la Argentina (1844–1912)* (Buenos Aires: Eudeba, 2004).

11 Hasia R. Diner, '"The most Irish city in the Union": the era of the Great Migration, 1844–1877', in Ronald H. Bayr and Timothy J. Meagher, eds., *The New York Irish* (Baltimore, MD: The Johns Hopkins University Press, 1996), pp. 87–106.
12 Ernest A. McKay, *The Civil War and New York City* (Syracuse, NY: Syracuse University Press, 1990), pp. 296–7, 308–9.
13 Edward O'Donnell, '"The scattered debris of the Irish nation": the famine Irish and New York City', in E. Margaret Crawford, ed., *The Hungry Stream: Essays on Emigration and Famine* (Omagh: Ulster-American Folk Park, Centre for Emigration Studies and Belfast: Queen's University, Institute of Irish Studies, 1997), pp. 53–54. This was in marked contrast to the pre-Famine immigrants who, though socially heterogeneous, tended to have a higher level of skills and be rather more prosperous than those who followed later; see Miller, *Emigrants and Exiles*, p. 295.
14 O'Donnell, '"The scattered debris of the Irish nation"', p. 54.
15 Quoted in Miller, *Emigrants and Exiles*, p. 323.
16 Browne, 'Government immigration policy in imperial Brazil, 1822–1870', p. 278.
17 Quote in Griggs, *The Elusive Eden*, p. 16.
18 On the complex post-war attitudes in the former Confederacy towards New York, and on the post-war migration from the South to the city, see Daniel E. Sutherland, *The Confederate Carpetbaggers* (Baton Rouge: Louisiana State University Press, 1988).
19 During the 1860s the Brazilian Legation was based in New York City, the Brazilian Minister only periodically visiting Washington, for a time situated on the front line in the war between Union and Confederate forces. Griggs, *The Elusive Eden*, pp. 29, 41, 165 n. 2; Quintino Bocaiúva, 'Uma questão pessoal: carta a Illustrada Redação de "A Província de São Paulo"', in Eduardo Silva, ed., *Idéias políticas de Quintino Bocaiúva* (Rio de Janeiro: Fundação Casa de Rui Barbosa, 1986), pp. 424–5; Blanche Henry Clark Weaver, 'Confederate emigration to Brazil', *The Journal of Southern History* 27/1 (1961), 41–2.
20 Laurence F. Hill, *Diplomatic Relations Between the United States and Brazil* (Durham, NC: Duke University Press, 1932), p. 244.
21 Griggs, *The Elusive Eden*, p. 83; *Anglo-Brazilian Times*, 22 April 1870; Weaver, 'Confederate emigration to Brazil', 44–5. The *Deutsche Zeitung*, a German-language newspaper published in Porto Alegre, later claimed that "several thousand" such people were recruited by Bocaiúva's agency, a figure that is almost certainly baseless and greatly inflated; see Bocaiúva, 'Uma questão pessoal', p. 425.
22 Browne, 'Government immigration policy in imperial Brazil, 1822–1870', pp. 147–8.
23 *Emigration to Brazil* (New York: S. Hallet, 1866), pp. 6–7.
24 *New York Tribune*, 30 November 1866, quoted in Griggs, *The Elusive Eden*, p. 41.
25 *New Orleans Times*, 10 February 1867, quoted in Griggs, *The Elusive Eden*, p. 42.
26 Hill, *Diplomatic Relations Between the United States and Brazil*, p. 243.
27 Griggs, *The Elusive Eden*, p. 121.
28 *Ibid*.
29 Mulhall, *Rio Grande do Sul and its German Colonies*, pp. 146–7.
30 D.M. Fox (Superintendent of the São Paulo Railway) to Mathew (Rio de Janeiro), 10 July 1872, NA/PRO FO128/100/165; Ian Lee (Anglican Chaplain in São Paulo) to Foreign Office, 29 May 1873, NA/PRO FO128/101/108. The national make-up of the immigrants is not entirely clear although Irish backgrounds predominated;

see Appendix 2.
31 Roger Swift, 'The historiography of the Irish in nineteenth-century Britain', in Patrick O'Sullivan, ed., *The Irish in the New Communities* (Leicester: Leicester University Press, 1992), p. 72.
32 Patrick O'Sullivan (Wednesbury, personal communication, 15 January 2001). Father O'Sullivan has examined the baptismal registers of St Mary's Church (Wednesbury) to determine the place of origin of the town's Irish immigrants. The surnames of 'heads of families' listed in the petition reproduced as Appendix 1 are further evidence of the Galway and Mayo origins of the Wednesbury Irish.
33 *The Rev. G. Montgomery's Register*, vol. 1, no. 5, 28 September 1867 (hereafter referred to as *Montgomery's Register*).
34 George J. Barnsby, *Social Conditions in the Black Country, 1800–1900* (Wolverhampton: Integrated Publishing Services, 1980), pp. 2–3 and Elihu Burritt, *Walks in the Black Country and its Green Border-land* (London: Sampson, Low, Son, and Marston, 1868), p. 351.
35 Burritt, *Walks in the Black Country*, p. 3.
36 Society for Promoting Christian Knowledge, *Birmingham and the Black Country* (London: Society for Promoting Christian Knowledge, 1864), p. 12.
37 Joseph Gillow, *A Literary and Biographical History, or Bibliographical Dictionary of the English Catholics*, vol. 5, (London: Burns & Oates, 1902), p. 87; Frederick William Hackwood, *Religious Wednesbury: Its Creeds, Churches & Chapels* (Dudley: The Herald Press, 1900), p. 93; Hugh Heinrick, *A Survey of the Irish in England (1872)* (London: The Hambledon Press, 1990), p. 46.
38 *The Weekly Register & Catholic Standard*, 19 March 1871.
39 *The Wednesbury and West Bromwich Advertiser*, 18 March 1871; *The Weekly Register & Catholic Standard*, 19 March 1871.
40 The first issue (vol. 1, no. 1) of *The Rev. G. Montgomery's Register* is dated 31 August 1867. The only known surviving copies of the *Register* are held by the Birmingham Archdiocesan Archives, St Chad's Cathedral, Birmingham (ref. P303/6/2), with the last number in the collection (vol. 1, no. 13) dated 4 July 1868; issue no. 7 is missing.
41 *Montgomery's Register*, vol. 1, no. 6, 19 October 1867. The entire text of the petition, along with the names of signatories, is reproduced as Appendix 1.
42 *Ibid.*
43 Heinrick, *A Survey of the Irish in England*, p. 49. Although Heinrick's report for *The Nation* was dated 20 August 1872, conditions would have been largely unchanged from just a few years earlier.
44 John F. Ede, *History of Wednesbury*, (Wednesbury: Wednesbury Corporation, 1962), p. 311.
45 Walter L. Arnstein 'The Murphy riots: a Victorian dilemma', *Victorian Studies* 19 (1975), 51–71.
46 *The Nation*, 6 June 1868, quoted in Heinrick, *A Survey of the Irish in England*, p. xi.
47 Ede, *History of Wednesbury*, pp. 317–8.
48 *Montgomery's Register*, vol. 1, no. 6, 19 October 1867.
49 Henry Formby, *A Voice from the Grave: Being the Funeral Discourse of the Rev. George Montgomery* (London: Burns, Oates, and Company, 1871), pp. 11–4.
50 Although there have been no studies focusing on onward migration of Irish people resident in England, conditions generally would indicate that the desire to move elsewhere was widespread. See David Fitzpatrick, *Oceans of Consolation: Personal Accounts of Irish Migration to Australia* (Cork: Cork University Press, 1994), pp. 334–58 for the frustrated experiences of one Irish family in the late nineteenth century trapped in Bolton by poverty, never quite able to scrape together the means to join relatives in far-off Queensland. A study relating to Bradford points to a steady

out-flow to the United States of Irish and other immigrant workers during the 1860s, with frequently one family member going on ahead with the intention of raising sufficient funds to send for those left behind; see Shelagh Mary Ward, 'Pennies of the poor: Catholics and poverty in Bradford, 1860–1914', unpublished MA thesis, Victorian Studies, Trinity and All Saints College, University of Leeds, 2000, pp. 12–3. On seasonal migrants in the early 19th century who, soon after returning to Ireland from England, migrated to the United States, see Ruth-Ann Harris, *The Nearest Place That Wasn't Ireland: Early Nineteenth-Century Labor Migration* (Ames: Iowa State University Press, 1994), pp. 185–8.

51 *Montgomery's Register*, vol. 1, no. 6, 19 October 1867 (emphasis in the original).
52 *Montgomery's Register*, vol. 1, no. 1, 31 August 1867; see also extracts of letters from contented Wednesbury emigrants in *Montgomery's Register*, vol. 1, no. 2, 7 September 1867 and in *Montgomery's Register*, vol. 1, no. 3, 14 September 1867; Formby, *A Voice from the Grave*, pp. 10–11.
53 *Montgomery's Register*, vol. 1, no. 5, 28 September 1867 (emphasis in the original).
54 *Montgomery's Register*, vol. 1, no. 1, 31 August 1867.
55 *Ibid.*
56 *Montgomery's Register*, vol. 1, no. 4, 21 September 1867; *Montgomery's Register*, vol. 1, no. 1, 31 August 1867.
57 Heinrick, *A Survey of the Irish in England*, p. 50.
58 Formby, *A Voice from the Grave*, pp. 10–11.
59 *Montgomery's Register*, vol. 1, no. 1, 31 August 1867 (emphasis in the original).
60 *Montgomery's Register*, vol. 1, no. 4, 21 September 1867.
61 See pp. 41–42 of this volume.
62 *Montgomery's Register*, vol. 1, no. 5, 28 September 1867, quoting from the 6 April 1866 edition of the *Standard*. The author of the article in the *Standard* was anonymous.
63 *Ibid.*
64 John Pascoe Grenfell (Brazilian consul, Liverpool) to Edward Pascoe Grenfell (Cardiff), 6 February 1868, Grenfell Papers, Corrêa do Lago Collection.
65 *The Brazil and River Plate Mail*, 2 November 1867.
66 Murdoch (Emigration Board, Colonial Office) to Rogers (Foreign Office), 15 February 1868, NA/PRO FO13/459; *The Anglo-Brazilian Times*, 23 March 1868.
67 Today called 'Florianópolis'; the contemporary name 'Desterro' will be used throughout.
68 One milreis was worth the equivalent of roughly 2s. 1d (two English shillings and a penny). One pound sterling consisted of twenty shillings.
69 *Montgomery's Register*, vol. 1, no. 8, 7 March 1868; *Montgomery's Register*, vol. 1, no. 5, 28 September 1867.
70 Brazil, *Relatório do Ministério e Secretario de Estado dos Negocios da Agricultura, Commercio e Obras Publicas* (1868), p. 22; letter from John Pascoe Grenfell (Brazilian consul, Liverpool) to Edward Pascoe Grenfell (Cardiff), 6 February 1868, Grenfell Papers, Corrêa do Lago collection; *Montgomery's Register*, vol. 1, no. 8, 7 March 1868; *Montgomery's Register*, vol. 1, no. 5, 28 September 1867; letter from Montgomery in *The Universal News*, 22 February 1868; letter from Montgomery in *The Universal News*, 2 June 1868. Montgomery placed weekly advertisements in *The Universal News* for most of 1868 appealing for financial support. Debts were not uncommon amongst Catholic clergy in England as the Church still faced financial restrictions and personal loans were often taken out by priests and bishops.
71 Letter from Montgomery in *The Universal News*, 22 February 1868; letter from Montgomery in *The Universal News*, 29 February 1868.
72 Letter from Montgomery in *The Universal News*, 22 February 1868; *The Brazil and River Plate Mail*, 7 February 1868.

73 *The Anglo-Brazilian Times*, 8 May 1868; *Montgomery's Register*, vol. 1, no. 8, 7 March 1868; *Birmingham Daily Post*, 4 February 1868; *The Universal News*, 15 February 1868; letter from Montgomery in *The Universal News*, 22 February 1868; Letter from Brazilian Legation (London) in *The Morning Post*, 25 February 1868.

74 *The Morning Post*, 24 February 1868; *The Liverpool Journal of Commerce*, 24 February 1868; *The Anglo-Brazilian Times*, 23 March 1868; Johann Jacob Sturz, *Die deutsche Auswanderung und die Verschleppung deutscher Auswanderer* (Berlin: Kortkampf, 1868), p. XXIII; Grenfell (Liverpool) to Brazilian Legation (London), 14 February 1868, Itamaraty, 217/4/1 Londres Oficíos Reservados (1865–1868), Stanley to Egas Moniz de Aragão (London), 15 February 1868, Itamaraty 217/4/1 Londres Oficios Reservados (1865–1868); Brazilian Chargé d'Affairs (London) to Foreign Office (London) 19 February 1868, NA/PRO FO13/459.

75 *The Anglo-Brazilian Times*, 9 December 1867.

76 Brazilian Chargé d'Affairs (London) to Foreign Office (London), NA/PRO FO13/459 19 February 1868; Murdoch (Colonial Office, London) to Foreign Office (London), 15 February 1868, NA/PRO FO13/459; Letter from Grenfell (Liverpool) to Brazilian Legation (London), 14 February 1868, Itamaraty, 217/4/1 Londres Oficios Reservados (1865–1868), Letter from Egas Moniz de Aragão (London) to notes on reply to Lord Stanley's verbal meeting, Itamaraty, 217/4/1 Londres Oficios Reservados (1865–1868).

77 Letter from Joaquim de Almeida Portugal, in *The Daily News*, 20 January 1868; *The Times*, 17 January 1868; *Birmingham Daily Post*, 4 February 1868.

78 *The Anglo-Brazilian Times*, 9 December 1867; *The Anglo-Brazilian Times*, 23 March 1868.

79 *The Anglo-Brazilian Times*, 23 March 1868.

80 Letter from Grenfell (Liverpool) to Brazilian Legation (London), 14 February 1868 Itamaraty 217/4/1 Londres Oficios Reservados (1865–1868); Prior (Emigration Officer, Liverpool) to Foreign Office (London), 14 February 1868, NA/PRO FO13/459.

81 *The Morning Post*, 24 February 1868; *The Liverpool Journal of Commerce*, 24 February 1868. Scully agreed later that rumours that British and German immigrants would be recruited for the army were utterly without foundation, *The Anglo-Brazilian Times*, 23 March 1868.

82 Careaga (São Vicente) to Wilkes (Foreign Office), 12 November 1868, NA/PRO FO13/459; Carlos Ficker, *Colonos de Joinville na guerra do Paraguai* (Blumenau: Blumenau em Cadernos, 1966).

83 Connolly (*Florence Chipman*, River Mersey) to Montgomery (Wednesbury), 24 February 1868, in *The Universal News*, 29 February 1868.

84 *The Universal News*, 15 February 1868. Bruce Chatwin commented of the leaders of the Welsh colonization efforts in Patagonia, that they "had combed the earth for a stretch of open country uncontaminated by Englishmen"; Bruce Chatwin, *In Patagonia* London: Jonathan Cape, 1977), p. 21. Although the first party of Welsh pioneers had only left for Patagonia in 1865, the scheme had been under discussion since the 1850s and by 1868 it was well known in both Wales and England; see Glyn Williams, *The Desert and the Dream: A Study of Welsh Colonization in Chubut, 1865–1915* (Cardiff: University of Wales Press, 1975), pp. 22–38. While Irish-speakers had long emigrated, the Irish language rarely established itself abroad. Most of the Irish who settled in Argentina came from the strongly Anglicised counties of Wexford and Westmeath although the Irish language was spoken by immigrants and their descendants from County Clare in an isolated pocket of the province of Buenos Aires, far away from the main areas of Irish immigration; see McKenna, 'Irish society in nineteenth century Argentina', p. 91, n. 24.

85 Prior (Government Emigration Office, Liverpool) to Walcot (Foreign Office, London)

20 February 1868, NA/PRO FO; *The Morning Post*, 24 February 1868; *The Anglo-Brazilian Times*, 23 March 1868. An original copy of the passport is with the Grenfell Papers, Corrêa do Lago Collection, but only a very partial listing of the emigrants survives; a copy is in the possession of the author.
86 *The Universal News*, 15 February 1868; *The Morning Post*, 24 February 1868.
87 *Montgomery's Register*, vol. 1, no. 9, 21 March 1868.
88 Letter from Montgomery in *The Universal News*, 22 February 1868; *The Universal News*, 15 February 1868.

3 A "New Ireland" in Brazil

1 *The Liverpool Journal of Commerce*, 24 February 1868; *Daily News*, 24 February 1868; Letter from Connolly (aboard the *Florence Chipman*, River Mersey) to Montgomery (Wednesbury), 24 February 1868, in *The Universal News*, 29 February 1868.
2 Letter from Montgomery (Wednesbury) dated 2 June 1868, *The Universal News* 6 June 1868.
3 Richard F. Burton, *Explorations of the Highlands of Brazil*, vol. 1, (London: Tinsley Brothers, 1869), p. 22.
4 Letter from Montgomery (Wednesbury) dated 2 June 1868, in *The Universal News* 6 June 1868; Hadfield, *Brazil and the River Plate*, p. 71.
5 *The Anglo-Brazilian Times*, 8 May 1868.
6 *The Anglo-Brazilian Times*, 23 April 1868; *The Brazil and River Plate Mail*, 22 May 1868.
7 Sturz, *Die deutsche Auswanderung und die Verschleppung deutscher Auswanderer*, pp. XXII–XXIII.
8 Letter from Montgomery (Wednesbury) dated 2 June 1868, *The Universal News* 6 June 1868; *Montgomery's Register*, vol. 1, no. 13, 4 July 1868.
9 *The Anglo-Brazilian Times*, 8 May 1868.
10 *The Anglo-Brazilian Times*, 23 March 1868; *The Anglo-Brazilian Times*, 23 April 1868; *The Anglo-Brazilian Times*, 8 May 1868 (emphasis in the original).
11 *The Anglo-Brazilian Times*, 8 May 1868.
12 *The Anglo-Brazilian Times*, 5 September 1868; *The Anglo-Brazilian Times*, 7 November 1868; Albert Jansen (Cananéia), July 1869, NA/PRO FO128/92.
13 John Hemming, *Amazon Frontier: The Defeat of the Brazilian Indians* (London: Macmillan, 1987), pp. 434–45. On Xokleng resistance to European land settlement efforts and the measures taken to annihilate the Indians, see also Sílvio Coelho dos Santos, *Indios e brancos no sul do Brasil: a dramática experiência dos Xokleng* (Porto Alegre: Editora Movimento,1987).
14 For a survey of early colonization schemes in Santa Catarina, see Maria Luiza Renaux Hering, *Colonização e indústria no Vale do Itajaí: o modelo catarinense de desenvolvimento* (Blumenau: Editora da Fundação Universidade de Blumenau, 1987), pp. 24–38 and Giralda Seyferth, *A colonização alemão no Vale do Itajaí-Mirim: um estudo de desenvolvimento econômico* (Porto Alegre: Editora Movimento, 1999), pp. 29–33.
15 Charles D'Ursel, *Sud-Amérique: séjours et voyages au Brésil, a la Plata, au Chili, en Bolivie et au Pérou* (Paris: E. Plon, 1879), pp, 177–8. On the Belgian experience, see Carlos Ficker, *Charles van Lede e a colonização belga em Santa Catarina. Subsídios para a história de colonização de Ilhota, no rio Itajaí-Açu, pela 'Compagne belge-brésilienne de colonisation'* (Blumenau: Blumenau em Cadernos, 1972).
16 Hering, *Colonização e indústria no Vale do Itajaí*, pp. 24–38; Seyferth, *A colonização alemão no Vale do Itajaí-Mirim*, pp. 29–33.

17 Seyferth, *A colonização alemão no Vale do Itajaí-Mirim*, pp. 41–3.
18 *The Anglo-Brazilian Times*, 7 August 1867; *The Anglo-Brazilian Times*, 22 April 1870.
19 *The Anglo-Brazilian Times*, 7 August 1867.
20 *Die Kolonie Zeitung*, 2 November, 1867; letter from Montgomery, *The Universal News*, 6 June 1868; *The Anglo-Brazilian Times*, 8 May 1868.
21 *The Anglo-Brazilian Times*, 22 April 1870.
22 *The Anglo-Brazilian Times*, 22 April 1870; Rebecca Cottle to Surgeon General, United States Army, 5 January 1889, Rousseau Family Papers, NC1040/1, University of Nevada Library, Special Collections, Reno.
23 *The Anglo-Brazilian Times*, 22 April 1870; Brazil, *Relatório do Ministério e Secretario de Estado dos Negocios da Agricultura, Commercio e Obras Publicas* (Rio de Janeiro, 1870), p. 39.
24 Brazil, *Relatório do Ministério e Secretario de Estado dos Negocios da Agricultura* (1870), p. 39; *The Anglo-Brazilian Times*, 7 August 1867; *Anglo-Brazilian Times*, 22 April 1870.
25 *Die Kolonie Zeitung*, 2 November 1867; *The Anglo-Brazilian Times*, 7 August 1867; *The Anglo-Brazilian Times*, 22 April 1870; quote from Aloisius Carlos Lauth, *A Colônia Príncipe Dom Pedro: um caso de política no Brasil Império* (Brusque: Privately published, 1987), pp. 21–4.
26 Santa Catarina, *Relatórios apresentados á Assembléa Legislativa Provincial de Santa Catarina, na sua sessão ordinaria, e ao Vice-Presidente Commendador Francisco José de Oliveira, no anno de 1868* (Rio de Janeiro, 1868), pp. 17–18.
27 *The Anglo-Brazilian Times*, 7 August 1867; *The Anglo-Brazilian Times*, 22 April 1870; Santa Catarina, *Relatórios apresentados á Assembléa Legislativa Provincial*, 1868, p. 18; Lauth, *A Colônia Príncipe Dom Pedro*, pp. 21–4.
28 *Kolonie Zeitung*, 30 March 1867; *The Anglo-Brazilian Times*, 7 August 1867; *The Anglo-Brazilian Times*, 22 April 1870; Lauth, *A Colônia Príncipe Dom Pedro*, pp. 21–5.
29 *Kolonie Zeitung*, 30 March 1867; *The Anglo-Brazilian Times*, 7 August 1867; *The Anglo-Brazilian Times*, 23 August 1867; *The Anglo-Brazilian Times*, 22 April 1870; Lauth, *A Colônia Príncipe Dom Pedro*, pp. 25–6. Galvão also investigated problems of Colônia Itajaí, in particular addressing demands for speedy distribution of title to land holdings. So angry were the *colonos* that soldiers were called to restore order at a demonstration staged in Itajaí; see Giralda Seyferth, *Colonização e conflito: estudo sobre "motins" e "desordens" numa região colonial de Santa Catarina no século XIX* (Rio de Janeiro: Museu Nacional – UFRJ Programa de Pos-graduação em Antropologia Social, Comunicação no. 10, 1988), p.36.
30 *Die Kolonie Zeitung*, 19 October 1867.
31 Lauth, *A Colônia Príncipe Dom Pedro*, p. 45; Sturz, *Die deutsche Auswanderung und die Verschleppung deutscher Auswanderer*, pp. IXL–XL; letter from Lazenby (Príncipe Dom Pedro) to Montgomery (Wednesbury), 3 January 1868, *The Universal News*, 29 February 1868; *Die Kolonie Zeitung*, 7 November, 1868.
32 Lauth, *A Colônia Príncipe Dom Pedro*, p. 45; Sturz, *Die deutsche Auswanderung und die Verschleppung deutscher Auswanderer*, pp. IXL–XL; letter from Lazenby (Príncipe Dom Pedro) to Montgomery (Wednesbury), 3 January 1868, *The Universal News*, 29 February 1868; *Die Kolonie Zeitung*, 7 November, 1868.
33 Lauth, *A Colônia Príncipe Dom Pedro*, p. 33; letter from Lazenby (Príncipe Dom Pedro) to Montgomery (Wednesbury), 3 January 1868, *The Universal News*, 29 February 1868; *Die Kolonie Zeitung*, 19 October 1867; *Die Kolonie Zeitung*, 7 November, 1868.
34 Santa Catarina, *Relatórios apresentados á Assembléa Legislativa Provincial*, 1868, p. 18.

35 *The Anglo-Brazilian Times*, 22 April 1870; *The Times*, 17 January 1868; Rebecca Cottle (Eureka, Nevada) to British Legation (Rio de Janeiro), 23 April, 1882, NA/PRO FO128/129; Corbett (Rio de Janeiro) to Rebecca Cottle (Eureka, Nevada), 13 July, 1882, NA/PRO FO128/129.
36 *Die Kolonie Zeitung*, 18 July 1868; Lauth, *A Colônia Príncipe Dom Pedro*, pp. 57–8.
37 *The Anglo-Brazilian Times*, 22 April 1870; Mathew (Rio de Janeiro) to Barão de Cotegipe (Rio de Janeiro), 10 April 1870, NA/PRO FO128/94.
38 *Die Kolonie Zeitung*, 23 May 1868; *Die Kolonie Zeitung*, 30 May 1868; Rogers (Colonial Office) to Foreign Office, 22 March 1870, NA/PRO CO386/119.
39 Bushby (Santos) to Cobbold (London), 12 July 1870, NA/PRO FO128/95; Haher (Príncipe Dom Pedro) to Watson (Desterro), 19 July 1869, NA/PRO FO128/90/175; *The Anglo-Brazilian Times*, 5 September 1868; *The Anglo-Brazilian Times*, 7 November 1868; *The Anglo-Brazilian Times*, 22 April 1870. Haher's background is not altogether clear, but it appears that he was recruited by the Brazilian Legation in London, arriving in Príncipe Dom Pedro at the end of January 1869: the salary that he was contracted to receive soon ceased; Brazil, *Relatório do Ministério e Secretario de Estado dos Negocios da Agricultura*, (1870), p. 40; Lauth, *A Colônia Príncipe Dom Pedro*, p. 83.
40 *The Anglo-Brazilian Times*, 5 September 1868.
41 Haher (Príncipe Dom Pedro) to Watson (Desterro), 19 July 1869, NA FO128/90/175.
42 *Die Kolonie Zeitung*, 9 May 1868.
43 Report from Príncipe Dom Pedro, 29 November 1868, *Die Kolonie Zeitung*, 2 January 1869; Elpídio de Mello (Rio de Janeiro) to Austin Orgill (Sharpsburgh, Pennsylvania), 24 June 1870, in Helen Jean Gent collection, Pennsylvania (copy in author's possession).
44 Report from Príncipe Dom Pedro, 29 November 1868, *Die Kolonie Zeitung*, 2 January 1869.
45 *Die Kolonie Zeitung*, 3 April 1869; Lauth, A Colônia Principe Dom Pedro, pp. 57–8; Haher (Príncipe Dom Pedro) to Watson (Desterro), 19 July 1869, NA/PRO FO128/90/175.
46 NA/PRO FO128/92/95.
47 Quoted in Seyferth, *Colonização e conflito*, p. 41.
48 Lauth, *A Colônia Príncipe Dom Pedro*, p. 70.
49 Watson (Desterro) to Callender (Rio Grande), 23 March 1869, NA/PRO FO128/90/172–4; Haher (Príncipe Dom Pedro) to Watson (Desterro), 19 July 1869, NA/PRO FO 128/90/175; Callender (Rio Grande) to Mathew (Rio de Janeiro) 21 August 1869, NA/PRO FO 128/90/178–80; undated (though apparently January/February 1869) draft of a petition to the President of Santa Catarina explaining the *colonos'* grievances written by Austin Orgill, in private collection, Pennsylvania (copy in the author's possession); *Die Kolonie Zeitung*, 11 September 1869.
50 Hunt (Rio de Janeiro) to Mathew (Rio de Janeiro), 15 June 1869, NA/PRO FO128/92.
51 Hunt (Rio de Janeiro) to Foreign Office (London), 15 June 1869, NA/PRO FO128/92; *The Anglo-Brazilian Times*, 23 June 1869.
52 Wright (Santos) to Fish (Washington) 17 May 1872, United States, Department of State, Dispatches from the United States Consuls in Santos, 1831–1906, vol. 2, (Microcopy T-351).
53 Mathew (Rio de Janeiro) to Leyard (London) 7 April 1874, British Library (Department of Manuscripts, Add. 39005, f. 215.
54 Lucian Barnsley (Tatuhy, São Paulo) to Godfrey Barnsley (Bartow County, Georgia), 27 April, 1870, Barnsley Papers, Perkins Library, Duke University. The New York recruits who had been sent to Nova Petrópolis or Caseiros in Rio Grande

do Sul suffered similarly – and were viewed with equal contempt. It was claimed that they bartered their food rations for liquor and used benches, doors and fences of the cabins where they were sheltered as firewood. Virtually all soon fled the settlements, making their way to Uruguay or to Brazilian cities; Mulhall, *Rio Grande do Sul and its German Colonies*, pp. 146–7.
55 Bushby (Santos) to Cobbold (London), 12 July 1870, NA/PRO FO128/95.
56 Hayes to *The Anglo Brazilian Times*, December 1869, NA/PRO FO128/92. This report entitled 'Some of the reasons why the Colonists of Príncipe Dom Pedro are not satisfied', was not published in *The Anglo-Brazilian Times*, although a copy was sent to the newspaper by Peter D. Hayes and to the British Consulate in Rio de Janeiro. It is not clear who Peter Hayes was but, on the basis of his extremely well-written report, he was clearly well educated and therefore not one of the New York or Wednesbury Irish. He may have been one of the relatively prosperous English immigrants who chose to stay on in Príncipe Dom Pedro.
57 Report from Príncipe Dom Pedro, 29 November 1868, *Die Kolonie Zeitung*, 2 January 1869; *Die Kolonie Zeitung*, 3 April 1869.
58 Hayes to *The Anglo Brazilian Times*, December 1869, NA/PRO FO128/92.
59 *Ibid.*
60 Undated (though probably December 1869) petition, in private collection, Pennsylvania (copy in the author's possession; spelling as in the original).
61 Martha Connolly Orgill (Sharpsburgh, Pennsylvania) to Mary Connerly (Liverpool), 29 May 1870 [spelling as in the original letter], Helen Jean Gent collection, Pennsylvania (copy in the author's possession). On 17 February Martha gave birth to another son, Thomas; Martha Connolly Orgill (Pittsburgh) to Mary Connerly (Liverpool), 27 February 1871, Helen Jean Gent collection, Pennsylvania (copy in the author's possession; spelling as in the original).
62 Elpídio de Mello (Rio de Janeiro) to Austin Orgill (Sharpsburgh, Pennsylvania), 24 June 1870, Helen Jean Gent collection, Pennsylvania (copy in the author's possession).
63 Mathew (Rio de Janeiro) to Cotegipe (Rio de Janeiro), 10 April 1870, NA/PRO FO128/94. A list, compiled in January 1870, includes the names of 146 British immigrants – 20 family units – resident in Príncipe Dom Pedro, some of whom would have left soon after; Mathew (Rio de Janeiro) to Cotegipe (Rio de Janeiro), 21 July 1870, NA/PRO FO128/94. Elpídio de Mello, who served as Director of Príncipe Dom Pedro from May to September 1868, wrote to a former *colono* who left Rio de Janeiro in February 1870 saying "I am really glad that you and your wife left this place already, for you don't know how difficult it is to receive now, from the Brazilian Government, any more transportations to the United States, or to any place out of Brazil, the chance you had, will never appear again", Elpídio de Mello (Rio de Janeiro) to Austin Orgill (Sharpsburgh, Pennsylvania), 24 June 1870, Helen Jean Gent collection, Pennsylvania (copy in the author's possession).
64 Bushby (Desterro) to Mathew (Rio de Janeiro) 23 May 1870, NA/PRO FO128/95; Bushby (Desterro) to Mathew (Rio de Janeiro) 6 June, 1870, NA/PRO FO128/95.
65 *Ibid.*; undated petition to British consul (Rio de Janeiro) signed by 24 Irish and US immigrants, NA/PRO FO128/95/375.
66 Bushby (Santos) to Cobbold (Rio de Janeiro), 30 June 1870, NA/PRO FO128/95; Charles Dundas (Santos) to Cobbold (Rio de Janeiro), 24 July 1870, NA/PRO FO128/95.
67 Bushby (Santos) to Cobbold (Rio de Janeiro), 12 July 1870, NA/PRO FO128/95.
68 Brazil, *Relatório do Ministério e Secretario de Estado dos Negocios da Agricultura*, (1870) pp. 40–1.
69 Letter from Revd J.I. Lee (English Church, São Paulo) dated 27 November 1872, *South American Missionary Magazine*, vol. VII, 1 March 1873.

70 J.F. Bromfield, *St. Mary's Centenary Souvenir*, no pagination; Ede, *History of Wednesbury*, p. 318; *O Dezenove de Dezembro*, 11 January 1871.
71 It is not clear at what point George Montgomery abandoned thoughts of travelling to Brazil, but it was clear that for sometime that his health was in a precarious state. By 1870 Montgomery had succumbed to "a painful illness" and on 7 March 1871 he died in Wednesbury. *The Dudley Herald and Wednesbury Borough News*, 11 March 1871; *The Wednesbury and West Bromwich Advertiser*, 18 March 1871; *The Weekly Register & Catholic Standard*, 19 March 1871; Formby, *A Voice from the Grave*.

4 Agricultural Labourers in Mid-Victorian England

1 Great Britain, *Report Respecting the Condition of British Emigrants in Brazil* (London: Parliamentary Paper, vol. 76, no. 3, 1874), p. 4.
2 For discussions on English rural life in the nineteenth century, see Pamela Horn, *Labouring Life in the Victorian Countryside* (Dublin: Gill and Macmillan, 1976); Gordon E. Mingay, *Rural Life in Victorian England* (Stroud: Alan Sutton, 1990); G.E. Mingay, ed., *The Victorian Countryside*, 2 vols, (London: Routledge & Kegan Paul, 1981).
3 Eric J. Hobsbawm and George Rudé, *Captain Swing* (London: Lawrence & Wishart, 1969), p. 52.
4 *The Illustrated London News*, 13 April 1872.
5 Nigel Scotland, *Agricultural Trade Unionism in Gloucestershire, 1872–1950*, (Cheltenham: The Centre for the Study of Religion, Cheltenham and Gloucester College of Higher Education, 1991), p. 15.
6 Elihu Burritt, *A Walk from London to Land's End and Back* (London: Sampson, Low, Son, and Marston, 1865), p. 118–9.
7 Hardy, *Tess of the D'Urbervilles*, p. 443.
8 Mingay, *Rural Life in Victorian England*, pp. 78–83.
9 *Ibid.*, pp. 78–84.
10 *Ibid.*, p. 78.
11 Frederick Law Olmsted, *Walks and Talks of an American Farmer in England* (Ann Arbor: Michigan University Press, 1969), p. 238.
12 Francis G. Heath, *British Rural Life and Labour* (London: P.S. King & Son, 1911), p. 230.
13 Francis G. Heath *The English Peasantry* (London: Frederick Warne and Co., 1874), pp. 47–48.
14 F.E. Green, *A History of the English Agricultural Labourer, 1870–1920* (London: P.S. King & Son, 1920), p. 37.
15 On the agricultural labourers' movement, see J.P.D. Dunbabin, 'The "Revolt of the Field": the agricultural labourers' movement in the 1870s', *Past and Present* 26 (1963), 68–98; Reg Groves, *Sharpen the Sickle! The History of the Farm Workers' Union* (London: The Merlin Press, 1981).
16 Burritt, *A Walk from London to Land's End and Back*, p. 434; Joseph Arch, *Joseph Arch: The Story of His Life, Told by Himself* (London: Hutchinson, 1898), p. 232.
17 See Dunbabin, '"The Revolt in the Field"', 68–98.
18 Scotland, *Agricultural Trade Unionism in Gloucestershire*, pp. 19–21.
19 *Ibid.*, p. 22.
20 Yeats quoted in Scotland, *Agricultural Trade Unionism in Gloucestershire*, p. 32.
21 *The Gloucester Journal*, 22 June 1872.
22 *The Gloucester Journal*, 6 July 1872.
23 *The Illustrated London News*, 27 April 1872; Dunbabin, 'The "Revolt of the Field"', 70.

24 *The Illustrated London News*, 13 April 1872.
25 Rule Book, cited in Scotland, *Agricultural Trade Unionism in Gloucestershire*, p. 17, n. 16.
26 Heath, *The English Peasantry*, p. 155.
27 The lower figure is generally considered the more likely of the two – see Pamela Horn, 'Agricultural trade unionism and emigration, 1872–1881', The Historical Journal 15 (1972) 97; the higher, based on Joseph Arch's own claims, includes 50,000 NALU-assisted emigrants between 1872 and 1874 alone – Groves, *Sharpen the Sickle!*, p. 67.
28 Charles Whitehead, *Agricultural Labourers* (London: Longmans, Green, Reader, and Dyer, 1870), p. 24.
29 NALU was not the first English organization representing agricultural labourers' interests to promote emigration, but it was the most persistent, assisting many thousands of its members. Established in 1871 as the first of this new wave of unions, the North Hertfordshire and South Shropshire Agricultural Labourers' Improvement Society had as its guiding motto "Emigration, migration, but not strikes". In addition to providing assistance to surplus labourers to migrate to northern England, the Society "emigrated" about forty people to North America. See Arch, *Joseph Arch*, p. 110.
30 Arch, *Joseph Arch*, p. 73.
31 On Minnesota, see *The Labourers' Union Chronicle*, 28 September 1872; on Texas, see *The Labourers' Union Chronicle*, September 1872; on Virginia, see *The Labourers' Union Chronicle*, 1 February 1873; and on Louisiana, see *The Labourers' Union Chronicle*, 27 September 1873.
32 Arch, *Joseph Arch*, p. 200.
33 *Ibid.*, p. 200.
34 See Horn, 'Agricultural trade unionism and emigration, 1872–1881', 87–102; John R. Millburn and Keith Jarrott, T*he Aylesbury Agitator. Edward Richardson, Labourers' Friend and Queensland Agent, 1849–1878* (Aylesbury: Buckinghamshire County Library, 1988); Rollo Arnold, *The Farthest Promised Land: English Villagers, New Zealand Immigrants of the 1870s* (Wellington: Victoria University Press, 1981); Tony Simpson, *The Immigrants: The Great Migration from Britain to New Zealand, 1830–1890* (Auckland: Godwit Publishing, 1997), pp. 155–87; Brian Birch, 'Popularizing the Plains: news of Kansas in England, 1860–1880', *Kansas History* 10/4 (1987/88), 262–74; John L. Harnsberger and Robert P. Wilkins, 'New Yeovil, Minnesota: a Northern Pacific colony in 1873', *Arizona and the West* 12/1 (1970), 5–22; Barbara J. Rozek, *Come to Texas: Enticing Immigrants, 1865–1915* (College Station, TX: Texas A&M University Press, 2003); on Canada, see Arch, *Joseph Arch*, pp. 174–98, 202.
35 *The Labourers' Union Chronicle*, 28 September 1872.
36 'Under the British Flag', *The Labourers' Union Chronicle*, 20 March 1875.

5 "The Workhouse – or Brazil": The Recruitment of English Emigrants

1 *The Labourers' Union Chronicle*, July 1872.
2 *The Royal Leamington Spa Courier and Warwickshire Standard*, 4 January 1873; *The Times*, 2 June 1873. Later investigations into the recruitment of immigrants referred to the system whereby agents were paid: "The plan of endeavouring to draw Emigration towards Brazil by vicious system of head-money cannot be too strongly condemned. It established a reward to avarice which to some men is a sore temptation, and to many an irresistible one. It opens a field to speculators which is too often strewn with the lives of misery of immigrant families. It introduces into the

country a class of people worthless in an economical sense, and dangerous in a social point of view. [....] Nor does the system of state colonies seem to me much better, [....] discredit[ing] the country abroad". O'Connor (Rio de Janeiro) to Mathew (Rio de Janeiro), 11 December 1876, NA/PRO FO128/109/463.
3 *The Royal Leamington Spa Courier and Warwickshire Standard*, 28 December 1872; E. Constantine Phipps (Rio de Janeiro), 4 January 1873, Great Britain, *Reports Respecting the Condition of British Emigrants in Brazil* (London: Parliamentary Paper, vol.75, 1873), p. 6.
4 *The Labourers' Union Chronicle*, September 1872; Great Britain, *Reports Respecting the Condition of British Emigrants in Brazil* (1873), p. 6).
5 *The Illustrated London News*, 27 April 1872.
6 L.F. Salzman, ed., *The Victoria History of the County of Warwick*, vol. VI, (London: Oxford University Press, 1951), p. 181.
7 *The Royal Leamington, Warwickshire, & Centre of England Chronicle*, 27 April 1872 (hereafter cited as the *Leamington Chronicle*). The NALU had in its Warwickshire heartland an enthusiastic champion in the *Leamington Chronicle*, supporting striking rural labourers and giving considerable positive publicity to emigration schemes to which the union was linked.
8 *Leamington Chronicle*, 27 April 1872; the basic terms were also to appear in advertisements placed by Brazil's Liverpool consulate in each issue of *The Labourers' Union Chronicle* from July 1872.
9 'A Traveller', *Leamington Chronicle*, 20 April 1872; *Leamington Chronicle*, 27 April 1872.
10 Great Britain, *Reports Respecting the Condition of British Emigrants in Brazil* (1873), pp. 72, 75–6, 81, 86–7.
11 *Leamington Chronicle*, 27 April 1872. For a copy of the prospectus see the *Labourers' Union Chronicle*, July 1872; or Great Britain, *Reports Respecting the Condition of British Emigrants in Brazil* (1873), appendix 3. Facilities to be expected were clearly drawn from the Brazilian Ministry of Agriculture's *Regulations of the State Colonies* of 19 January 1867.
12 Great Britain, *Reports Respecting the Condition of British Emigrants in Brazil* (1873), p. 20.
13 William Ebenezer Yeats, quoted in Scotland, *Agricultural Trade Unionism in Gloucestershire*, p. 30.
14 Great Britain, *Reports Respecting the Condition of British Emigrants in Brazil* (1873), p. 62.
15 Dundas (Paranaguá) to Mathew (Rio de Janeiro), 30 June 1874, NA/PRO FO128/104/143. Consul Hunt who saw a copy of the leaflet in Paranaguá, found it entirely inaccurate, the real prices, for example, being almost twice those stated.
16 'A Traveller', *Leamington Chronicle*, 20 April 1872. The unidentified author of the letter with this comment was not, however, much more complimentary of other destinations open to immigrants, also warning against Queensland (lack of rain and rivers and too expensive to reach) and Canada (winters too long and too cold). Only Texas was looked upon favourably owing to its supposedly good climate, high wages and possibilities of acquiring land.
17 Great Britain, *Reports Respecting the Condition of British Emigrants in Brazil* (1873), p. 62.
18 Olmsted, *Walks and Talks of an American Farmer in England*, p. 238.
19 Statement by Louisa Bayliss, James Millard, Reuben Walker, Albert Estcourt and Albert Lane (Mendes) 16 February 1873 enclosed with Mathew (Rio de Janeiro) to Granville (London), 21 February 1873, NA/PRO FO128/101/300.
20 Mathew (Rio de Janeiro) to Granville (London), 15 January 1873, NA/PRO FO128/101/279.

21 Great Britain, *Reports Respecting the Condition of British Emigrants in Brazil* (1873), pp. 86–7.
22 *Leamington Chronicle*, 20 April 1872.
23 *Ibid.*
24 *The Banbury Guardian*, 2 January 1873. The "cabbage" reference is, presumably, to the cabbage palmetto, a common species of palm featuring edible white leaf buds, the layers of which vaguely resemble a cabbage in appearance.
25 John Crossley (Assunguy), 15 September 1872, *Leamington Chronicle*, 30 November 1872.
26 Edward Lyons (Cananéia), 17 August 1872, *The Labourers' Union Chronicle*, 21 December 1872 (also appearing in *The Times*, 24 December 1872).
27 *The Labourers' Union Chronicle*, 4 January 1873.
28 *Ibid.* Emphasis in the original.
29 *The Times*, 12 April 1873.
30 Letter from Edward Lyons (Cananéia), 17 August 1872, *The Labourers' Union Chronicle*, 21 December 1872 (also appearing in *The Times*, 24 December 1872).
31 J.W. Apperley (Assunguy), 6 November 1872, *The Labourers' Union Chronicle*, 4 January 1873.
32 *Ibid.*
33 *Ibid.*
34 *The Royal Leamington Spa Courier and Warwickshire Standard*, 11 January 1873.
35 Hunt (Rio de Janeiro) to Foreign Office (London), 8 October 1874, NA/PRO FO128/102/497 commenting on the *The Labourers' Union Chronicle*, 4 January 1873.
36 Letter from Edward Lyons (Cananéia), 17 August 1872, *The Labourers' Union Chronicle*, 21 December 1872.
37 Compare, for example, with letters in Erickson, *Invisible Immigrants*; Fitzpatrick, *Oceans of Consolation;* and Josephine Wtulich, ed., *Writing Home: Immigrants in Brazil and the United States, 1890–1891* (Boulder, CO: East European Monographs, 1986).
38 *The Banbury Guardian* (no date given), cited in John R. Hodgkins, *Over the Hills to Glory: Radicalism in Banburyshire, 1832–1945* (Southend: Clifton Press, 1978), p. 69.

6 From England to the Brazilian Colonies

1 *The Labourers' Union Chronicle*, June 1872; *The Labourers' Union Chronicle*; July 1872; letter from Frederick Payn (14 July 1872), *The Banbury Guardian*, 20 March 1873; Thomas P. Bigg-Wither, *Pioneering in South Brazil* vol. 1, (London: John Murray, 1878) p. 6.
2 Letter from Thomas Alsop (Rio de Janeiro, 1 July 1872), *Labourers Union Chronicle*, September 1872; letter from Thomas Alsop, *Labourers Union Chronicle*, 2 November 1872; letter from Frederick Payn (14 July 1872), *Banbury Guardian*, 20 March 1873; Bigg-Wither, *Pioneering in Southern Brazil*, vol. 1, p. 6–14.
3 Letter from Thomas Alsop, *The Labourers' Union Chronicle*, 2 November 1872.
4 *The Labourers' Union Chronicle*, June 1872; *The Rugby Advertiser*, 18 May 1872.
5 Letter from Thomas Alsop (Leamington), *The Times*, 14 April 1873.
6 Letter from Thomas Alsop (Rio de Janeiro, 1 July 1872), *The Labourers' Union Chronicle*, September 1872; letter from Frederick Payn (14 July 1872), *The Banbury Guardian*, 20 March 1873; letter from Thomas Alsop, *The Labourers' Union Chronicle*, 2 November 1872. It is not known how many British immigrants were transferred to Montebello. Twenty-three men (some accompanied by wives and children) were certainly recruited, with the following individuals listed as having settled

there, no matter how briefly: Thomas Bell, Henry Burgain, Henry Duke, Richard Farmer, William Hawes, John Hornby, Robert Hornby, Henry Brough Jones, John J. Mead, James Oakes, Thomas Owens, Francis Parker, Frederick Payne, Michael Smith, William Smith, John Taylor, John Taylor, Jr., George Tew, George Thompson, Henry Thompson, Edward Tudor, Thomas Wilkinson and Walter Williams. Contracts dated 28 June 1872 and 2 October 1872, Montebello, NA/PRO FO128/100/128.

7 Letter from Thomas Alsop (Rio de Janeiro, 1 July 1872), *The Labourers' Union Chronicle*, September 1872.
8 Letter from Frederick Payn (14 July 1872), *The Banbury Guardian*, 20 March 1873.
9 Letter from Thomas Alsop, *The Labourers' Union Chronicle*, 2 November 1872.
10 Letter from Thomas Alsop, *The Labourers' Union Chronicle*, 2 November 1872; letter from Thomas Alsop (Leamington), *The Times*, 14 April 1873.
11 Letter from Thomas Alsop, *The Labourers' Union Chronicle*, 2 November 1872.
12 Letter from Thomas Alsop (Rio de Janeiro, 1 July 1872), *The Labourers' Union Chronicle*, September 1872.
13 Great Britain, *Reports Respecting the Condition of British Emigrants in Brazil* (1873), p. 23.
14 'Relatório da Agência Official de Colonisação, in Brazil' in Brazil, *Relatório do Ministério e Secretario de Estado dos Negocios da Agricultura, Commercio e Obras Publicas* (1872), Annexo D, p. 8.
15 Letter from Thomas Alsop, *The Labourers' Union Chronicle*, 2 November 1872.
16 Great Britain, *Reports Respecting the Condition of British Emigrants in Brazil* (1873) p. 26.
17 Letter from Thomas Alsop (1 July 1872), *The Labourers' Union Chronicle*, September 1872.
18 Letter from Thomas Alsop, *The Labourers' Union Chronicle*, 2 November 1872.
19 *Ibid.*
20 *Ibid.*; letter from Thomas Alsop, *The Labourers' Union Chronicle*, 16 November 1872.
21 *The Labourers' Union Chronicle*, 16 November 1872. The claim is not entirely fanciful, Dom Pedro having visited during a European tour the annual Royal Agricultural Show in Warwickshire, one of the most important events of its kind in Britain.
22 *The Labourers' Union Chronicle*, 16 November 1872; *The Labourers' Union Chronicle*, 23 November 1872.
23 Statements by John Nottley and John Gillingham of Piddletown, Dorset, in Great Britain, *Report Respecting the Condition of British Emigrants in Brazil* (1874), p. 7.
24 Letter from William Brown (Cananéia, 28 July 1872), *The Royal Leamington Spa Courier and Warwickshire Standard*, 23 November 1872, a few days earlier published in *The Banbury Guardian*.
25 Letter from Mary Stanton (Cananéia, 23 August 1872), *The Labourers' Union Chronicle*, 21 December 1872. The letter also appeared in *The Royal Leamington Spa Courier and Warwickshire Standard*, 23 November 1872.
26 Letter from Thomas Alsop (Leamington), *The Royal Leamington Spa Courier and Warwickshire Standard*, 21 December 1872.
27 Letter from Thomas Alsop (Leamington), *The Times*, 14 April 1873.
28 Great Britain, *Reports Respecting the Condition of British Emigrants in Brazil* (1873), pp. 5–6.
29 Letter from Thomas Alsop (Leamington), *The Times*, 15 April 1873.
30 Petition of 26 "heads of family" (Rio de Janeiro) to Phipps (British Legation, Rio de Janeiro), February 1873, NA/PRO FO/128/101/319.
31 Statements signed by returned immigrants from Cananéia (Mendes), 16 February 1873, NA/PRO FO128/101/302.
32 Statements signed by returned immigrants from Cananéia (Mendes), 16 February

1873, NA/PRO FO128/101/302.
33 Statements signed by returned immigrants from Cananéia (Mendes), 16 February 1873, NA/PRO FO128/101/302; Phipps (Petrópolis), 19 March 1873 to British immigrants at Mendes, NA/PRO FO128/100/260.
34 Information on the Bond family in Cirencester obtained from the Marriage, Births and Deaths registers and the 1861 and 1871 censuses for England, Family History Centre. Thanks are due to Mauro Fortes Carneiro, of Curitiba, for helping to piece together the story of the Bond family.
35 Statement by Clara Bond (Paranaguá), 2 June 1874, NA/PRO FO128/104/165.
36 Petition of 26 "heads of family" (Rio de Janeiro) to Phipps (British Legation, Rio de Janeiro), February 1873, NA/PRO FO128/101/319; letter from Antonina, 26 July 1873, *The Wilts and Gloucestershire Standard*, 13 September 1873; statement by Clara Bond (Paranaguá), 2 June 1874, NA/PRO FO128/104/165; *O Dezenove de Dezembro*, 19 July 1873; Dundas (Paranaguá) to Mathew (Rio de Janeiro), 30 June 1874, NA/PRO FO128/104/143; deposition by William Ryan (Paranaguá), 3 June 1874, insert in Dundas (Rio de Janeiro) to Mathew (Rio de Janeiro), July 1874, NA/PRO FO128/104/161; Phipps (Rio de Janeiro, 17 February) to Mathew (Rio de Janeiro), in Great Britain, *Reports Respecting the Condition of British Emigrants in Brazil* (1873), p. 67.
37 Report from Rio de Janeiro, 26 February 1873, NA128/99/278; Phipps (Petrópolis), 19 March 1873, NA/PRO FO128/100/260.
38 Copy of Yates & Co. prospectus and contract (1872) in NA/PRO FO128/100/122–9; Thomas Dutton (Rio de Janeiro) to Mathew (Rio de Janeiro), 13 March 1873, NA/PRO FO128/100/130.
39 Petition of 26 "heads of family" (Rio de Janeiro) to Phipps (British Legation, Rio de Janeiro), February 1873, NA/PRO FO/128/101/319; letter from Antonina, 26 July 1873, *The Wilts and Gloucestershire Standard*, 13 September 1873; statement by Clara Bond (Paranaguá), 2 June 1874, NA/PRO FO128/104/165; *O Dezenove de Dezembro*, 19 July 1873; Dundas (Paranaguá) to Mathew (Rio de Janeiro), 30 June 1874, NA/PRO FO128/104/143; deposition by William Ryan (Paranaguá), 3 June 1874, insert in Dundas (Rio de Janeiro) to Mathew (Rio de Janeiro), July 1874, NA/PRO FO18/104/161; Phipps (Rio de Janeiro, 17 February) to Mathew (Rio de Janeiro), in Great Britain, *Reports Respecting the Condition of British Emigrants in Brazil* (1873), p. 67.
40 *Ibid.*; petition of 26 "heads of family" (Rio de Janeiro) to Phipps (British Legation, Rio de Janeiro), February 1873, NA/PRO FO128/101/319; Roberto Edgar Lamb, U*ma jornada civilizadora: imigração, conflíto social e segurança pública na Província do Paraná, 1867 a 1882* (Curitiba: Aos Quatro Ventos, 1997), pp.11–3.
41 Bigg-Wither, *Pioneering in South Brazil*, vol. 2, pp. 188–9; petition of 26 "heads of family" (Rio de Janeiro) to Phipps (British Legation, Rio de Janeiro), February 1873, NA/PRO FO128/101/319.
42 Statement by James Hutchings (Paranaguá), 31 May 1874, NA/PRO FO128/104/163.
43 Statement by Clara Bond (Paranaguá), 5 May 1874, NA/PRO FO128/105/91; Statement by Clara Bond (Paranaguá), 2 June 1874, NA/PRO FO128/104/165.
44 Letter from Jane Landner (Antonina, 26 July 1874), *The Wilts and Gloucestershire Standard*, 13 September 1874.
45 *The Wilts and Gloucestershire Standard*, 4 October 1873; Davis, a railway worker first in Gloucester and then Cirencester, had travelled to Brazil with his wife and child where they both died.
46 *Ibid.*; George Harmer (*The Wilts and Gloucestershire Standard*, Cirencester) to Granville (London), 12 September 1873, NA/PRO FO128/101/187; *The Wilts and Gloucestershire Standard*, 27 September 1873; *The Wilts and Gloucestershire Standard*,

22 November 1873.
47 Letter from the Foreign Office (20 November 1873), *The Wilts and Gloucestershire Standard*, 29 November 1873. A further letter from the Foreign Office (21 January 1874) in *The Wilts and Gloucestershire* of 24 January 1874 reported that the Brazilian government had informed the British minister in Rio that the Bond children were all well cared for and living in Curitiba, Charles having been placed with Dr João Jose Pedrosa; Anna, with Lieutenant-Colonel Cândido Jose Pereira; Ernest, with Captain Antonio Enes Bandeiras; Alfred, with Antonio Marcel de Oliveira; and Emma, with Lieutenant Gabriel de Almeida Torres. Later that year, *The Anglo-Brazilian Times* of 7 August 1874 reiterated that the children were being cared for by "humane people".
48 *The Wilts and Gloucestershire Standard*, 22 November 1873, quoting an article that appeared in the *Anglo-Brazilian Times*
49 *Ibid.* (emphasis added).
50 *O Dezenove de Dezembro*, 19 July 1873; *O Dezenove de Dezembro*, 23 July 1873.
51 *O Dezenove de Dezembro*, 23 July 1873; *O Dezenove de Dezembro*, 30 July 1873. See also Lamb, *Uma jornada civilizadora*, pp. 13–5.
52 *O Dezenove de Dezembro*, 23 July 1873; *O Dezenove de Dezembro*, 30 July 1873; Dundas (Paranaguá) to Mathew (Rio de Janeiro), 30 June 1874, NA/PRO FO128/ 104/143.
53 Letter sent from Assunguy (10 August 1873), *O Dezenove de Dezembro*, 21 September 1873.
54 *The Wilts and Gloucestershire Standard*, 4 October 1873.
55 Tigar (Assunguy) to Preston (Rio de Janeiro), 25 February 1874, NA/PRO FO128/105/70.
56 Deposition by James Hutchings (Paranaguá), 5 May 1874, NA/PRO FO128/105/86; statement by Clara Bond (Paranaguá), 2 June 1874, NA/PRO FO128/104/165. According to later English census returns, Clara Bond somehow succeeded in obtaining a passage back to England and she and baby Clara returned to Cirencester. It is not known what became of James Hutchings.

7 Pioneering in South Brazil

1 Murdoch to Standford, 17 December, 1869, NA/PRO FO128/93/3; Brazil, *Relatório do Ministério e Secretario de Estado dos Negocios da Agricultura, Commercio e Obras Publicas* (1868), p. 43.
2 *The Times*, 8 February 1868; *The Daily Telegraph*, 25 February 1868; Brazil, *Relatório do Ministério e Secretario de Estado dos Negocios da Agricultura, Commercio e Obras Publicas* (1868), p. 43.
3 Murdoch (Emigration Board, London) to Sandford (Foreign Office), 17 December 1869, NA/PRO FO128/93/3; *The Universal News*, 15 February 1868; *The Times*, 17 January 1868; *The Anglo-Brazilian Times*, 5 September 1868; petition signed by 21 British male immigrants (Assunguy) to Mathew (Rio de Janeiro), 1 March 1869, NA/PRO FO128/92/260; Bigg-Wither, *Pioneering in South Brazil*, vol. 2, pp. 180–1; Caroline Tamplin (Assunguy), 14 September 1875, NA/PRO FO128/108/286.
4 Murdoch (Emigration Board, London) to Sandford (Foreign Office), 17 December 1869, NA/PRO FO128/93/3; *The Universal News*, 15 February 1868; *The Times*, 17 January 1868; *The Anglo-Brazilian Times*, 5 September 1868; petition signed by 21 British male immigrants (Assunguy) to Mathew (Rio de Janeiro), 1 March 1869, NA/PRO FO128/92/260.
5 Report from Ernest Buhlaw (Jacapiranga, 5 December 1865) *The Anglo-Brazilian Times*, 2 January 1866.
6 John Codman, *Ten Months in Brazil; with Notes on the Paraguayan War* (Edinburgh:

R. Grant & Son, London: Simpkin, Marshall & Co. and New York: D. Appleton & Co., 1870), pp. 181–2.
7   See Griggs, *The Elusive Eden*.
8   Frank P. Goldman, *Os pioneiros americanos no Brasil: educadores, sacerdotes, covos, e reis* (São Paulo: Livraria Pioneira Editora, 1972).
9   Codman, *Ten Months in Brazil*, pp. 181–2; report from Ernest Buhlaw (Jacapiranga), 5 December 1865, in *The Anglo-Brazilian Times*, 2 January 1866.
10 Letter from Ernest A. Buhlaw (Cananéia), 26 June 1866 to Mr Nathan, *The Anglo-Brazilian Times*, 7 August, 1866; report from Ernest Buhlaw (Jacapiranga, 5 December 1865) *The Anglo-Brazilian Times*, 2 January 1866; Great Britain, *Emigration to Brazil. Report on the Colony of Cananea* (London: Parliamentary Paper, vol.82, 1875); *Emigration to Brazil* (New York: S. Hallet, 1866), p. 48.
11 Great Britain, *Emigration to Brazil. Report on the Colony of Cananea*, pp. 4–5.
12 Great Britain, *Emigration to Brazil. Report on the Colony of Cananea*, pp. 6–7; James Davies (Cananéia) to Corbett (Rio de Janeiro), 10 May 1884, NA/PRO FO128/138; letter from E.W. Young (Cananéia, 10 November 1872), forwarded by William Hadfield, *The Times*, 14 April 1873; deposition of Joseph Kilchen (Rio de Janeiro), 1876, NA/PRO FO128/112/13; Lt. Commander C.G. May (HMS *Rifleman*, Rio de Janeiro), 2 October 1888, NA/PRO FO128/157; petition signed by eleven British male immigrants (Cananéia), 24 January 1870 to Mathew (Rio de Janeiro), NA/PRO FO128/94/294.
13 Report from Ernest Buhlaw (Jacapiranga, 5 December 1865) in *The Anglo-Brazilian Times*, 2 January 1866.
14 Report from Ernest Buhlaw (Jacapiranga, 5 December 1865) in *The Anglo-Brazilian Times*, 2 January 1866; petition signed by eleven British male immigrants (Cananéia), 24 January 1870 to Mathew (Rio de Janeiro), NA/PRO FO128/94/294; *Emigration to Brazil*, p. 48; Great Britain, *Emigration to Brazil. Report on the Colony of Cananea*, pp. 8–9.
15 Great Britain, *Emigration to Brazil. Report on the Colony of Cananea*, p. 4.
16 *The Anglo-Brazilian Times*, 9 April 1878.
17 Letter from Charles Young (Cananaéia) to William Hadfield, dated 16 July 1872, *The Labourers' Union Chronicle*, 19 October 1872.
18 Great Britain, *Emigration to Brazil. Report on the Colony of Cananea*, pp. 9–10; *The Anglo-Brazilian Times*, 9 April 1878.
19 Letter from Ernest A. Buhlaw (Cananéia), dated 26 June 1866, to Mr Nathan, *The Anglo-Brazilian Times*, 7 August, 1866; report from Ernest Buhlaw (Jacapiranga, 5 December 1865) *The Anglo-Brazilian Times*, 2 January 1866.
20 Great Britain, *Emigration to Brazil. Report on the Colony of Cananea*, p. 12; Deposition of Joseph Kilchen (Rio de Janeiro), 1876, NA/PRO FO128/112/13.
21 Deposition of Edward Lyons (Rio de Janeiro), 19 December 1876, NA/PRO FO128/112/21.
22 May, (HMS *Rifleman*, Rio de Janeiro), 2 October 1888, NA/PRO FO128/157; Great Britain, Emigration to Brazil. *Report on the Colony of Cananea*, p. 9–12 and 17; petition signed by eleven male British immigrants (Cananéia) to Mathew (Rio de Janeiro), 24 January 1870, NA/PRO FO128/94/294.
23 Great Britain, *Emigration to Brazil. Report on the Colony of Cananea*, pp. 15–6.
24 Petition signed by eleven British male immigrants (Cananéia), 24 January 1870 to Mathew (Rio de Janeiro), NA/PRO FO128/94/294; Great Britain, *Emigration to Brazil. Report on the Colony of Cananea*, p. 11; Albert Jansen (Cananéia), July 1869, FO128/92.
25 Honoraria Joyce (Cananéia) to British Minister (Rio de Janeiro), 17 January 1878, NA/PRO FO128/116.
26 Deposition of Joseph Kilchen (Rio de Janeiro), 1876, NA/PRO FO128/112/13;

Joyce (Cananéia) to British Minister (Rio de Janeiro), 17 January 1878, NA/PRO FO128/116.
27 Deposition of Joseph Kilchen (Rio de Janeiro), 1876, NA/PRO FO128/112/13.
28 Great Britain, *Emigration to Brazil. Report on the Colony of Cananea*, p. 15.
29 Deposition of Edward Lyons (Rio de Janeiro), 19 December 1876, NA/PRO FO128/112/21.
30 Great Britain, *Emigration to Brazil. Report on the Colony of Cananea*, pp. 15–6.
31 Petition dated 2 October 1868 addressed to the Minister of Agriculture and signed by Henry George Young, Edward Pedder, James Muller, James Hennessey, Jeremiah Collins, Daniel Daly, Stephen Hubbard, Richard Noonan, James Davies, Charles Hardiman, H. A. Levy, J. Barton, Albert Jansen, J. Mitchell, E. Mitchell, *The Anglo-Brazilian Times*, 7 November 1868.
32 Mathew (Petrópolis) to Cotepige (Rio de Janeiro), 10 April 1870, NA/PRO FO128/94/219.
33 Petition signed by eleven British male immigrants (Cananéia), 24 January 1870 to Mathew (Rio de Janeiro), NA/PRO FO128/94/294.
34 Great Britain, *Emigration to Brazil. Report on the Colony of Cananea*, p. 17.
35 Petition signed by eleven British male immigrants (Cananéia), 24 January 1870 to Mathew (Rio de Janeiro), NA/PRO FO128/94/294.
36 Dundas (Santos) to Mathew, 21 February 1874, NA FO128/104/106.
37 Great Britain, *Emigration to Brazil. Report on the Colony of Cananea*, pp. 19 and 23.
38 Great Britain, *Emigration to Brazil. Report on the Colony of Cananea*, pp. 22–4.
39 Great Britain, *Emigration to Brazil. Report on the Colony of Cananea*, pp. 19–22.
40 Thomas P. Bigg-Wither, *Pioneering in South Brazil*, vol. 2 (1878), pp. 170–8; M.I.P. Queiroz, 'Messiahs in Brazil', *Past and Present* 31 (1965), 37, 67–8; Rui Wachowicz, 'O comércio da madeira e a atuação da Brazil Railway no sul do Brasil', *Boletimdo Instituto Histórico, Geografico e Etnográfico Paranaense* 42 (1984), 51.
41 Bigg-Wither, *Pioneering in South Brazil*, vol. 2, pp. 170–8; Queiroz, 'Messiahs in Brazil', 37, 67–8; Wachowicz, 'O comércio da madeira e a atuação da Brazil Railway no sul do Brasil', 51.
42 Bigg-Wither, *Pioneering in South Brazil*, vol. 2, p. 174.
43 Wilson Martins, *Um Brasil diferente: ensai o sobre fenôminos da aculturação no Parané*, 2. ed., (São Paulo: T.A. Queiroz, 1989), p. 64.
44 Great Britain, *Emigration to Brazil. Report on the Colony of Assunguy* (London: Parliamentary Paper, vol. 82, 1875), p. 3
45 Great Britain, *Emigration to Brazil. Report on the Colony of Assunguy*, p. 11
46 Great Britain, *Emigration to Brazil. Report on the Colony of Assunguy*, p. 8.
47 Paraná, *Relatório apresentado a Assembléia Legistlativa do Paraná no dia 15 de Feberero de 1872 pelo Presidente da Provincia da o excellentissimo senhor Oliveira Lisboa* (Curitiba, 1872), p. 53.
48 Martins, *Um Brasil diferente*, p. 80; Paraná, *Relatório apresentado a Assembléia Legistlativa do Paraná no dia 15 de Feberero de 1872 pelo Presidente da Provincia da o excellentissimo senhor Oliveira Lisboa*, p. 53; 'Relatório da Agência Official de Colonisação, in Brazil', in Brazil, *Relatório do Ministério e Secretario de Estado dos Negocios da Agricultura, Commercio e Obras Publicas* (1872), Annexo D, p. 13–4.
49 Great Britain, *Reports Respecting Conditions of British Emigrants in Brazil* (1873), pp. 28 and 31.
50 Anonymous correspondent (Assunguy), *O Dezenove de Dezembro*, 16 February 1870.
51 *Ibid.*
52 The largest single party of so-called "Algerians" left Marseilles in late November 1868 on the French vessel the Polymeric and they arrived in Curitiba three months later. Brazil, *Relatório do Ministério e Secretario de Estado dos Negocios da Agri-*

cultura Commercio e Obras Publicas, (1870), p. 41; *O Dezenove de Dezembro*, 3 July 1869; *O Dezenove de Dezembro*, 7 July 1869; *O Dezenove de Dezembro*, 11 August 1869; 21 August 1869; *O Dezenove de Dezembro*, 16 October 1869; Mulhall, *Rio Grande do Sul and its German Colonies*, p. 196.
53 *O Dezenove de Dezembro*, 29 July 1874.
54 Bigg-Wither, *Pioneering in South Brazil*, vol. 2, p. 186.
55 Great Britain, *Emigration to Brazil. Report on the Colony of Assunguy*, pp. 2–3.
56 Bigg-Wither, *Pioneering in South Brazil*, vol. 2, pp. 167–8; Carlos Brent Cenci (Commissioner of the Brazilian government) to Ignâcio da Cunha Galvão (Official Agent of Colonization), 15 November 1873, NA/PRO FO128/49.
57 *Ibid.*
58 Bigg-Wither, *Pioneering in South Brazil*, vol. 2, pp. 167–8.
59 Bigg-Wither, *Pioneering in South Brazil*, vol. 2, p. 168–9, 191; Carlos Brent Cenci (Commissioner of the Brazilian government) to Ignâcio da Cunha Galvão (Official Agent of Colonization), 15 November 1873, NA/PRO FO128/49; Dundas (Santos) to Mathew, 21 February 1874, NA/PRO FO128/104/106; Dundas (Santos) to Mathew (Rio de Janeiro), 30 June 1874, NA/PRO FO128/104/143; Great Britain, *Emigration to Brazil. Report on the Colony of Assunguy*, pp. 4–5; Bangeo Balster, 'Report on Assunguy' (Paranaguá) to Dundas (Santos), 27 November 1875, NA/PR FO128/108/307; Tigar (Assunguy) to Drummond (Rio de Janeiro), 13 July 1875, NA/PRO FO128/105/402.
60 Bigg-Wither, *Pioneering in South Brazil*, vol. 2, p. 185.
61 Great Britain, *Emigration to Brazil. Report on the Colony of Assunguy*, p. 5; Bigg-Wither, *Pioneering in South Brazil*, vol. 2, p. 185; letter (Assunguy, 22 August 1873) *O Dezenove de Dezembro*, 3 September 1873.
62 Carlos Brent Cenci (Commissioner of the Brazilian government) to Ignâcio da Cunha Galvão (Official Agent of Colonization), 15 November 1873, NA/PRO FO128/49; Bigg-Wither, *Pioneering in South Brazil*, vol. 2, p. 185–6; director's report (15 February 1870), *O Dezenove de Dezembro*, 30 March 1870; letter (Assunguy, 22 August 1873) *O Dezenove de Dezembro*, 3 September 1873; PP, Brazil. No. 2 (1875): 6, 13–4; Great Britain, *Emigration to Brazil. Report on the Colony of Assunguy*, p. 13; Lee (Assunguy) to Drummond (Rio) 29 April 1875, NA FO128/105/389; Tigar (Assunguy) to Drummond (Rio de Janeiro) 13 July 1875, NA/PRO FO128/105/402.
63 Mathew to Visconde de Caravellas, 24 January 1874, NA/PRO FO128/103/18; Great Britain, *Emigration to Brazil. Report on the Colony of Assunguy*, pp. 6, 17–9; Dundas (Santos) to Mathew (Rio de Janeiro), 30 June 1874, NA/PRO FO128/104/143.
64 Great Britain, *Emigration to Brazil. Report on the Colony of Assunguy*, pp. 6–7; Bigg-Wither, *Pioneering in South Brazil*, vol. 2, pp. 180–1; Caroline Tamplin (Assunguy), 14 September 1875, NA/PRO FO128/108/286; O'Connor (Rio de Janeiro) to Mathew (Rio de Janeiro), 11 December 1876, NA/PRO FO128/109/463.
65 Great Britain, *Emigration to Brazil. Report on the Colony of Assunguy*, pp. 9–10.
66 *Ibid.*, pp. 6–8, 13; Bigg-Wither, *Pioneering in South Brazil*, vol. 2, pp. 180–1; Caroline Tamplin (Assunguy), 14 September 1875, NA/PRO FO128/108/286; O'Connor (Rio de Janeiro) to Mathew (Rio de Janeiro), 11 December 1876, NA/PRO FO128/109/463.
67 Deposition by William Ryan (Paranaguá), 26 April 1874, NA/PRO FO128/105/88; deposition by William Ryan (Paranaguá), 3 June 1874, insert in Dundas (Rio de Janeiro) to Mathew (Rio de Janeiro), July 1874, NA/PRO FO128/104/161.
68 Samuel Newport (Assunguy), 14 November 1875, NA/PRO FO128/108/305; Bangeo Balster, 'Report on Assunguy' (Paranaguá) to Dundas (Santos), 27 November 1875, NA/PR FO128/108/307; Tigar (Assunguy) to Drummond (Rio de Janeiro), 9 January 1876, NA/PRO FO128/108/289.
69 Frederick Tigar (Assunguy) to Revd Preston (Rio de Janeiro), 28 February 1874,

NA/PRO FO128/105/70.
70 Relatório do Director da Colonia do Assunguy (Manoel Barata Goéz), 30 Dezembro 1879, APP 0582 (Arquivo Público do Paraná) – Ofícios 1879 – v. 23, p. 154.
71 Relatório do Director da Colonia do Assunguy (Manoel Barata Goéz), 30 Dezembro 1879, APP 0582 – Ofícios 1879 – v. 23, p. 152.
72 Paraná, *Relatorio apresentado a Assemblea Legistlativa do Paraná no dia 15 de Feberero de 1876 pelo Presidente da Provincia da o excellentissimo senhor doutor Adolpho Lins* (Curitiba, 1876), p. 84; Relatório do Director da Colonia do Assunguy (Manoel Barata Goéz), 30 Dezembro 1879, APP 0582 – Ofícios 1879 – v. 23, pp. 150–1; *O Dezenove de Dezembro*, 11 January 1871; Samuel Newport (Assunguy), 14 November 1875, NA/PRO FO128/108/305; William D. Robinson (Assunguy), November 1875, NA/PRO FO128/108/305; Dundas (Santos) to Mathew (Rio de Janeiro), 30 June 1874, NA/PRO FO128/104/143; Bangeo Balster, 'Report on Assunguy' (Paranaguá) to Dundas (Santos), 27 November 1875, NA/PR FO128/108/307.
73 Tigar (Assunguy) to Revd Preston (Rio de Janeiro), 28 February 1874, NA/PRO FO128/105/70; Tigar (Assunguy) to Drummond (Rio de Janeiro), 13 July 1875, NA/PRO FO128/108/292.
74 Bangeo Balster, 'Report on Assunguy' (Paranaguá) to Dundas (Santos), 27 November 1875, NA/PR FO128/108/307.
75 *O Dezenove de Dezembro*, 11 January 1871; Dundas (Santos) to Mathew (Rio de Janeiro), 30 June 1874, NA/PRO FO128/104/143; Bangeo Balster, 'Report on Assunguy' (Paranaguá) to Dundas (Santos), 27 November 1875, NA/PR FO128/108/307; O'Connor (Rio de Janeiro) to Mathew (Rio de Janeiro), 11 December 1876, NA/PRO FO128/109/463.
76 O'Connor (Rio de Janeiro) to Mathew (Rio de Janeiro), 11 December 1876, NA/PRO FO128/109/463.
77 Tigar (Assunguy) to Drummond (Rio de Janeiro), 13 July 1875, NA FO128/105/402; John Lee, formerly British chaplain in São Paulo(Assunguy) to Drummond (Rio de Janeiro), 29 April 1875, NA/PRO FO128/105/389; Carlos Brent Cenci (Commissioner of the Brazilian government) to Ignâcio da Cunha Galvão (Official Agent of Colonization), 15 November 1873, NA/PRO FO128/49; Mathew to Visconde de Caravellas, 24 Jan 1874 NA/PRO FO128/102/496; William D. Robinson (Assunguy), November 1875, NA/PRO FO128/108/305.
78 Deposition by William Ryan (Paranaguá), 26 April 1874, NA/PRO FO128/105/88; Bigg-Wither, *Pioneering in South Brazil*, vol. 2, p. 185.
79 O'Connor (Rio de Janeiro) to Mathew (Rio de Janeiro), 11 December 1876, NA/PRO FO128/109/463.
80 Bangeo Balster, 'Report on Assunguy' (Paranaguá) to Dundas (Santos), 27 November 1875, NA/PR FO128/108/307; Mathew (Rio de Janeiro) to Minister of Agriculture (Rio de Janeiro), 20 December 1876, NA/PRO FO128/108/421.
81 Dundas (Santos) to Mathews (Rio de Janeiro), 30 June 1874, FO128/104/143; Great Britain, *Emigration to Brazil. Report on the Colony of Assunguy*, pp. 10–1, 15.
82 Tigar (Assunguy) to Drummond (Rio de Janeiro), 13 July 1875, NA/PRO FO128/108/292 (emphasis in the original).
83 Tigar (Assunguy) to Preston (Rio de Janeiro), 25 February 1874, NA/PRO FO128/105/70.
84 Tigar (Assunguy) to Drummond (Rio de Janeiro), 9 January 1876, NA/PRO FO128/108/289.
85 Great Britain, *Report Respecting the Condition of British Emigrants in Brazil* (1874), p. 3.
86 *O Dezenove de Dezembro*, 29 July 1874.

87 Paraná, *Relatorio apresentado a Assemblea Legistlativa do Paraná no dia 15 de Feberero de 1876 pelo Presidente da Provincia da o excellentissimo senhor doutor Adolpho Lins* (Curitiba, 1876).
88 Bigg-Wither, *Pioneering in South Brazil*, vol. 2, p. 186–7.
89 Tigar (Assunguy) to Drummond (Rio de Janeiro), 13 July 1875, NA/PRO FO128/105/402.
90 Tigar (Assunguy) to Preston (Rio de Janeiro), 25 February 1874, NA/PRO FO128/105/70; Tigar (Assunguy) to Drummond (Rio de Janeiro), 13 July 1875, NA/PRO FO128/108/292.
91 Caroline Tamplin (Assunguy), 14 September 1875, NA/PRO FO128/108/286; Tamplin (Curitiba), 18 September 1881, NA/PRO FO128/116.
92 Tamplin (Assunguy), 14 September 1875, NA/PRO FO128/108/286.
93 Great Britain, *Reports Respecting the Condition of British Emigrants in Brazil* (1873), pp. 29 and 47–8.
94 Redman (Assunguy) to British Legation (Rio de Janeiro), 22 January 1880, NA/PRO FO128/116; Harris Gastrell, Chargé d'Affairs (Petrópolis) to Granville, Foreign Office (London), 20 September 1880, NA/PRO FO128/121.
95 Deposition by William Ryan (Paranaguá), 3 June 1874, insert in Dundas (Rio de Janeiro) to Mathew (Rio de Janeiro), July 1874, NA/PRO FO128/104/161; *The Anglo-Brazilian Times*, 23 July 1874; Tigar (Assunguy) to Preston (Rio de Janeiro), 28 February 1874, NA/PRO FO128/105/70; Bangeo Balster, 'Report on Assunguy' (Paranaguá) to Dundas (Santos), 27 November 1875, NA/PR FO128/108/307.
96 Great Britain, *Emigration to Brazil. Report on the Colony of Assunguy*, p. 14; Tamplin (Assunguy) to Mathew (Rio de Janeiro), 6 March 1877, NA/PRO FO128/112/542.
97 Tamplin (Assunguy), 14 September 1875, NA/PRO FO128/108/286; Tigar (Assunguy) to Drummond (Rio de Janeiro), 13 July 1875, NA/PRO FO128/ 108/292; Bangeo Balster, 'Report on Assunguy' (Paranaguá) to Dundas (Santos), 27 November 1875, NA/PR FO128/108/307; O'Connor (Rio de Janeiro) to Mathew (Rio de Janeiro), 11 December 1876, NA/PRO FO128/109/463; Tamplin (Curitiba), 18 September 1881, NA/PRO FO128/116; Albert Tigar, untitled memoirs (Tigar family, Vancouver), p. 5.
98 Tamplin (Assunguy), 14 September 1875, NA/PRO FO128/108/286. Albert Tigar, an Assunguy-born grandson of Charles and Caroline Tamplin, recorded in his memoirs that, "my mother's father who as a practicing physician at St. Johns Ward [St John's Wood] outside of London [...] became possessed of the wanderlust and decided to sell off everything and migrate to Brazil and be attached to the Colony as a doctor". According to Albert, his grandfather retired as the colony's doctor due to ill-health. Albert Tigar, untitled memoirs, (Tigar family, Vancouver), pp. 5, 22.
99 Joseph Renaudin pursued (apparently unsuccessfully) his claim for back pay for years, asking the British Legation to intervene on his behalf. Renaudin (Assunguy) to Drummond (Rio de Janeiro), 13 July 1875, NA/PRO FO128/105/314; Renaudin (Assunguy) to Mathew (Rio de Janeiro), 12 May 1879, NA/PRO FO128/116; Renaudin (Assunguy) to Mathew (Rio de Janeiro), 25 December 1879, NA/PRO FO128/116.
100 *The Anglo-Brazilian Times*, 23 July 1874; Mathew to Caravellas, 24 January 1874, NA/PRO FO128/103/18.

8 The Collapse

1 Bangeo Balster, 'Report on Assunguy' (Paranaguá) to Dundas (Santos), 27 November 1875, NA/PR FO128/108/307.
2 Tigar (Assunguy) to Drummond (Rio de Janeiro), 13 July 1875, NA/PRO FO128/108/292.

Notes

3   Petition signed by 21 male British immigrants (Assunguy) to Mathew (Rio de Janeiro), 1 March 1869, NA/PRO FO128/92/260.
4   Petition signed by eleven British male immigrants (Cananéia), 24 January 1870 to Mathew (Rio de Janeiro), NA/PRO FO128/94/294.
5   William Robinson (Curitiba) to British Minister (Rio de Janeiro), 7 March 1881, NA/PRO FO128/116; Robinson (Curitiba) to British Minister (Rio de Janeiro), 26 June 1881, NA/PRO FO128/116;
6   Great Britain, *Emigration to Brazil. Report on the Colony of Cananea*, p. 17.
7   James Hurst (Rio de Janeiro) to Mathew (Rio de Janeiro), 6 February 1869, NA/PRO FO128/92/263.
8   *The Times*, 15 September 1873; *The Wilts and Gloucester Standard*, 20 September 1873.
9   Mathew (Rio de Janeiro) to Viscount de Carcovelles (Brazilian Foreign Ministry), 13 October 1873, NA FO128/98/506; Hunt (Rio de Janeiro) to Mathew (Petrópolis), 29 January 1874, FO128/104/78.
10 List of male "heads of families", wives and children, in Mathew (Petrópolis) to Granville (London), 22 January 1874, NA FO128/102/286; *The Anglo-Brazilian Times*, 22 January 1874; Tigar (Assunguy) to Preston (Rio de Janeiro), 28 February 1874, NA FO128/105/70.
11 Letter reproduced in English (with a Portuguese translation) *O Dezenove de Dezembro*, 11 February 1874.
12 Hunt (Rio de Janeiro) to Mathew (Petrópolis), 29 January 1874, NA/PRO FO128/104/78; *The Anglo-Brazilian Times*, 22 January 1874.
13 Unattributed report on Assunguy, October 1873, NA/PRO FO128/49/281–284.
14 *The Anglo-Brazilian Times*, 23 October 1874.
15 *The Anglo-Brazilian Times*, 23 October 1874.
16 Mathew (Rio de Janeiro) to Leyard (London), 7 April 1874, British Library, Add. 39005 f. 215; Mathew (Petrópolis) to Granville (London), 20 February 1874, NA/PRO FO128/102/310; Hunt (Rio de Janeiro) to Mathew (Petrópolis), 29 January 1874, FO128/104/78; *The Anglo-Brazilian Times*, 22 January 1874; Great Britain, *Reports Respecting the Condition of British Emigrants in Brazil* (1873), p.3; list of male "heads of families", wives and children, in Mathew (Petrópolis) to Granville (London), 22 January 1874, NA/PRO FO128/102/286.
17 Mathew (Rio de Janeiro) to Leyard (London), 7 April 1874, British Library, Add. 39005 f. 215.
18 Great Britain, *Report Respecting the Condition of British Emigrants in Brazil* (1874), p.1; George Preston (Chaplain, Rio de Janeiro) to Mathew (Rio de Janeiro), 6 February 1874, NA/PRO FO128/102/294.
19 Mathew (Petrópolis) to Granville (London), 20 February 1874, NA/PRO FO128/102/310.
20 Mathew (Rio de Janeiro) to Leyard (London), 7 April 1874, British Library, Add. 39005 f. 215.
21 Hunt (Rio de Janeiro) to Mathew (Petrópolis), 29 January 1874, NA/PRO FO128/104/78.
22 Granville (London) to Mathew (Rio de Janeiro), 20 March 1874, NA/PRO FO128/102/2.
23 For example, approval was given on 16 September 1874 to the passages of James Watts and children and on 21 October 1874 to the passages of the Mitchell and Weston families. Derby (London) to Mathew (Rio de Janeiro), 16 September 1874, NA/PRO F0128/102/188; Derby (London) to Mathew (Rio de Janeiro), NA/PRO FO128/102/232.
24 Derby (London) to Mathew (Rio de Janeiro), NA/PRO FO128/102/252.
25 Levy Walsh (Paranaguá) to Mathew (Rio de Janeiro), 1 May 1874, NA/PRO

FO128/105/85.
26 Dundas (Santos) to Mathew (Rio de Janeiro), 30 June 1874, NA/PRO FO128/104/143.
27 *Ibid.*
28 *Ibid.*
29 Deposition by Ellen Godwin (Paranaguá), 26 April 1874, NA/PRO FO128/105/88.
30 Deposition by Frank White (Paranaguá), May 1874, NA/PRO FO128/105/87.
31 Mathew (Petrópolis) to Dundas (Consul, Santos), 10 May 1874, NA/PRO FO128/104/19.
32 Drummond (Petrópolis) to Derby (London), 5 March 1875, NA/PRO FO128/106/382; Mathew (Rio de Janeiro) to Caravellas (Rio de Janeiro), 25 March 1874, NA/PRO FO128/103/38; Mathew (Rio de Janeiro) to Caravellas (Rio de Janeiro), 27 March 1874, NA/PRO FO128/103/39.
33 O'Connor (Rio de Janeiro) to Mathew (Rio de Janeiro), 11 December 1876, NA/PRO FO128/109/463.
34 Minister of Agriculture (Rio de Janeiro) to Mathew (Rio de Janeiro), 16 March 1877, NA/PRO FO128/111/97.
35 Joseph Redman (St John's Parsonage, Assunguy) to Mathew (Rio de Janeiro), NA/PRO FO128/112/599.
36 Redman (Assunguy) to British Legation (Rio de Janeiro), 28 June 1879, NA/PRO FO128/116.
37 Redman (Assunguy) to Mathew (Rio de Janeiro), 7 December 1877, NA/PRO FO128/112/607.
38 Redman (Assunguy) to Mathew (Rio de Janeiro), 28 October 1878, NA/PRO FO128/116; British Legation (Rio de Janeiro) to Redman (Assunguy), 23 November 1878, NA/PRO FO128/116; Redman (Assunguy) to Mathew (Rio de Janeiro), 28 December 1878, NA/PRO FO128/116.
39 Redman (Assunguy) to Edwin Corbett, British Ambassador (Rio de Janeiro), 18 November 1883, NA/PRO FO128/129.
40 Tigar (Assunguy) to Drummond (Rio de Janeiro), 13 July 1875, NA/PRO FO128/108/292.
41 *Ibid.*
42 Tigar (Assunguy) to Rev. John Lee (Colchester), 29 January 1879, NA/PRO FO128/117.
43 William Robinson (Assunguy), November 1875, NA/PRO FO128/108/305.
44 Tigar (Assunguy) to Lee (Colchester), 29 January 1879, NA/PRO FO128/117
45 Redman (Assunguy) to British Legation (Rio de Janeiro), 28 June 1879, NA/PRO FO128/116; Foreign Office (London) to Lee (Colchester), 20 August 1879, NA/PRO FO128/117; Redman (Assunguy) to J. Harris Gastrell, British Chargé d'Affairs (Rio de Janeiro), 6 October 1880, NA/PRO FO128/116; Tigar (Assunguy) to British Legation (Rio de Janeiro), 8 October 1880, NA/PRO FO128/116; Granville (London) to Gastrell (Rio de Janeiro) 28 October 1880, NA/PRO FO128/121.
46 Ford (Rio de Janeiro) to Granville (London), 12 November 1880, NA/PRO FO128/121. Frederick and Katherine's eldest son, Albert, was six years old when the family left Brazil. He left a manuscript describing, in remarkable detail, his family's experiences as pioneer farmers in Brazil and, from 1881, following a brief stay in London, in southwest Manitoba. Albert claimed that it was his father's ill-health – a fever "contracted in the West Indies in his early days" – that made the family decide to abandon an otherwise happy life in Brazil. This and other inconsistencies with contemporary documents limit the value of the Brazilian-portion of this manuscript as historical evidence, although it tells a fascinating story of two failed, but contrasting, experiences of pioneer settlement. Albert Tigar, untitled memoirs (Tigar family, Vancouver), pp. 36–7.

Notes

47 William Robinson (Curitiba) to British Minister (Rio de Janeiro), 7 March 1881, NA/PRO FO128/116; Robinson (Curitiba) to British Minister (Rio de Janeiro), 26 June 1881, NA/PRO FO128/116;
48 Tamplin (Curitiba), 18 September 1881, NA/PRO FO128/116. Albert Tigar, a grandson of Caroline, whose childhood memories of Brazil were probably mixed with the reminiscences of his parents, recorded that his grandmother left Assunguy, as in Curitiba she could have "more of the comforts of life and less of the hardships." Albert Tigar, untitled memoirs, (Tigar family, Vancouver), p. 22. Caroline Tamplin's diaries survive for the early 1880s (Tamplin family papers, Curitiba). The short daily entries detail her monotonous life, listing teaching and social engagements, church attendance, weather conditions, health and the music that she was playing and the books that she was reading. Unfortunately the only references to Assunguy are brief mentions of people travelling between Curitiba and the colony.
49 Redman (Assunguy) to Gastrell (Rio de Janeiro), 6 October 1880, NA/PRO FO128/116.
50 Petition signed by 105 British residents of Assungy to Mathew (Rio de Janeiro), 30 June 1879, NA/PRO FO128/116; Redman (Assunguy) to British Legation (Rio de Janeiro), 22 January 1880, NA/PRO FO128/116; Relatório do Director da Colonia do Assunguy (Manoel Barata Goéz), 30 Dezembro 1879, APP 0582 – Ofícios 1879 – v. 23, p. 163.
51 British Legation (Rio de Janeiro) to Salisbury (Foreign Office, London), 8 August 1879, NA/PRO FO128/117.
52 Redman (Assunguy) to Gastrell (Rio de Janeiro), 17 May 1880, NA/PRO FO128/116; Redman (Assunguy) to British Consulate (Rio de Janeiro), 5 June 1880, NA/PRO FO128/116.
53 Redman (Assunguy) to Gastrell (Rio de Janeiro), 18 August 1880, NA/PRO FO128/116.
54 Granville (London) to Gastrell (Rio de Janeiro), 18 June 1880, NA/PRO FO128/121.
55 Gastrell (Rio de Janeiro) to Redman (Assunguy), 8 October 1880, NA/PRO FO128/116; Ministério dos Negocios Estrangeiros (Rio de Janeiro) to British Legation (Rio de Janeiro), 10 March 1881, NA/PRO FO128/114/810; Ministério dos Negocios Estrangeiros (Rio de Janeiro) to British Legation (Rio de Janeiro), 6 June 1881, NA/PRO FO128/114/927; Redman (Assunguy) to Corbett (Rio de Janeiro), 18 November 1883, NA/PRO FO128/129; Redman (Assunguy) to Corbett (Rio de Janeiro), 18 September 1883, NA/PRO FO128/129.
56 Redman (Curitiba) to Corbett (Rio de Janeiro), 11 August 1883, NA/PRO FO128/129.
57 Redman (Assunguy) to Gastrell (Rio de Janeiro), 6 October 1880, NA/PRO FO128/116; Redman (Assunguy) to Ford, Chargé d'Affairs, (Rio de Janeiro), 28 January 1881, NA/PRO FO128/116;
58 Redman (Assunguy) to Corbett (Rio de Janeiro), 18 February 1882, NA/PRO FO128/129; Redman (Assunguy) to Corbett (Rio de Janeiro), 18 September 1883, NA/PRO FO128/129.
59 Corbett (Rio de Janeiro) to Cowper (Santos), 3 December 1883, NA/PRO FO128/129; Redman (Assunguy) to Corbett (Rio de Janeiro), 28 July 1883, NA/PRO FO128/129.
60 Redman (Curitiba) to Corbett (Rio de Janeiro), 11 August 1883, NA/PRO FO128/129; Redman (Assunguy) to Corbett (Rio de Janeiro), 18 September 1883, NA/PRO FO128/129.
61 Cooper, British consul (Santos) to Corbett (Rio de Janeiro), 4 February 1884, NA/PRO FO128/136.
62 Cooper (Santos) to Corbett (Rio de Janeiro), 4 February 1884, NA/PRO FO128/136; Royal Mail Steam Packet Company (Rio de Janeiro) to Corbett, 30 November 1883, NA/PRO FO128/136.
63 Redman (Desborough, Northamptonshire) to Corbett (Rio de Janeiro), 25 March 1884, NA/PRO FO128/138.

64 See, for example, letter from John Cameron (Cananéia) to Mathew (Rio de Janeiro), NA/PRO FO128/116; James Davies (Cananéia) to Corbett (Rio de Janeiro), 16 February 1884, NA/PRO FO128/138.
65 Cowper, British consul (Santos) to British Legation (Rio de Janeiro), 25 April 1887, NA/PRO FO128/150; Cowper (Santos) to British Legation (Rio de Janeiro), 3 June 1887, NA/PRO FO128/150.
66 James Davies (Cananéia) to Corbett (Rio de Janeiro), 10 May 1884, NA/PRO FO128/138.
67 Davies (Cananéia) to Corbett (Rio de Janeiro), 10 May 1884, NA/PRO FO128/138. See also Appendix 2 for a final letter from Bridget Davies.

Conclusion

1 Rogers (Colonial Office) to Foreign Office, 22 March 1870, NA/PRO CO386/119. The "vagabonds and ragamuffins" charge was made in the Porto Alegre German-language newspaper, the *Deutsche Zeitung*, criticizing those who recruited the New York Irish, instead of examining the failures of the Ministry of Agriculture and the colonization system in general; quoted by Bocaiúva 'Uma questão pessoal', p. 425.
2 Minister of Agriculture Joaquim Antão Leão was especially vitriolic. But Scully claimed that Leão was shocked by the effect budget cuts were having on state colonies and approved during his last days in office measures to enable the president of Santa Catarina to assist Príncipe Dom Pedro *colonos*. Scully added that Leão's replacement as minister of agriculture, Diego Velho de Cavalcante, granted a large additional payment to state colonies to resume halted public works; *The Anglo-Brazilian Times*, 22 April 1870; *The Anglo-Brazilian Times*, 23 December 1868.
3 *The Anglo-Brazilian Times*, 23 October 1868.
4 *The Labourers' Union Chronicle*, July 1872.
5 Comment by William Howitt, first published in *Anti-Game-Law Circular*, a radical newsletter, and reproduced in *The Wilts and Gloucester Standard*, 8 February 1873.
6 *The Wilts and Gloucester Standard*, 21 June 1873.
7 Letter from Thomas Alsop (Leamington), *The Royal Leamington Spa Courier and Warwickshire Standard*, 21 December 1872 (emphasis in the original).
8 Letter from 'Massa Sahib', *The Royal Leamington Spa Courier and Warwickshire Standard*, 28 December 1872.
9 Letter from Thomas Sheasby (Cananéia, 3 November 1872), *Royal Leamington Spa Courier and Warwickshire Standard*, 11 January 1873.
10 *Ibid.*
11 *The Royal Leamington Spa Courier and Warwickshire Standard*, 11 January 1873.
12 *The Royal Leamington Spa Courier and Warwickshire Standard*, 11 January 1873; *The Wilts and Gloucester Standard*, 22 February 1873.
13 Pownoll W. Phipps (Vicar of Napton), *The Times*, 29 September 1873.
14 Arch, *Joseph Arch*, p. 202.
15 By mid-1873, Yeats was in disagreement with the NALU leadership over his district's financial affairs and the union was all too willing to seize on the Brazilian fiasco further to discredit Yeats. In September, the NALU formally disowned Yeats; Scotland, *Agricultural Trade Unionism in Gloucestershire*, p. 32.
16 Letter signed "England for Ever", *The Royal Leamington Spa Courier and Warwickshire Standard*, 30 November 1872.
17 *The Wilts and Gloucester Standard*, 20 September 1873. New Zealand in particular was now being presented as "a Paradise for the Working Man" where, after just a few years of employment as a house or farm servant, an immigrant could save enough to

start a farm of his own. Prospects in the Pacific colony were said to be so good that even "idlers and drunkards" could find satisfaction there, as they "are able to earn by two or three days labour sufficient to keep them in food and in a state of semi-intoxication the rest of the week"; *The Labourers' Union Chronicle*, 13 September 1873.
18 *The Labourers' Union Chronicle*, 27 September 1873.
19 *The Times*, 11 April 1873.
20 *The Times*, 11 April 1873.
21 Bigg-Wither, *Pioneering in South Brazil*, vol. 2, pp. 186–7.
22 "Jacaré Assu", pseud., *Brazilian Colonization from an European Point of View* (London: Edward Stanford, 1873), p. 5.
23 *Ibid.*, pp. 7–8.
24 *Ibid.*, p.125.
25 Great Britain, *Report Respecting the Condition of British Emigrants in Brazil* (1873), p. 1.
26 Dundas (Santos) to Mathew (Rio de Janeiro), 30 June 1874, NA/PRO FO128/104/143; Dundas (Santos) to Mathew (Rio de Janeiro), 21 February 1874, NA/PRO FO128/104/106.
27 Great Britain, *Report Respecting the Condition of British Emigrants in Brazil* (1873), p. 2.
28 *Hansard's Parliamentary Debates*, 17 Feb 1872, vol. 214, pp. 532–6; *Hansard*, 9 May 1873, vol. 215, pp. 1712–3; *Hansard*, 23 June 1873, vol. 216, p. 1249.
29 *Hansard*, 17 Feb 1872, vol. 214, pp. 532–6.
30 Letter from Joaquim Fernando Antão Leão (Rio de Janeiro) dated 23 November, 1868, *The Anglo-Brazilian Times*, 8 December 1868.
31 Ministério dos Negocios Estrangeiros (Rio de Janeiro) to Drummond (Rio de Janeiro), 25 April 1876, NA/PRO FO128/108/75.
32 *Ibid.*
33 *Ibid.*
34 *O Dezenove de Dezembro*, 25 March 1874, quoting Relatório, 15 February 1874.
35 Report by J.S.A.C. Rosa dated 25 May 1874, *O Dezenove de Dezembro*, 13 June 1874.
36 Emílio Nunes Correia de Menezes, 17 January 1866, quoted in Martins, *Um Brasil diferente*, p. 12.
37 *O Dezenove de Dezembro*, June 1870; *O Dezenove de Dezembro*, 18 August 1875.
38 Relatório do Director da Colônia do Assunguy (Manoel Barata Goéz), 30 Dezembro 1879, APP 0582 – Ofícios 1879 – v. 23, p. 151.
39 Brazil, *Relatório do Ministério e Secretario de Estado dos Negocios da Agricultura Commercio e Obras Publicas* (1872), pp. 146–7.
40 Carlos Brent Cenci (Commissioner of the Brazilian government) to Ignâcio da Cunha Galvão (Official Agent of Colonization), 15 November 1873, NA/PRO FO128/49. The report was also not without criticism of the British immigrants. Although he accepted that some of the immigrants were "sober and industrious", he considered most as being "fault-finding, boisterous, discontented people, who would be dissatisfied with the most favourable circumstances; and who, in my opinion, will never be any benefit either to Brazil or themselves".
41 Paraná, *Relatório apresentado a Assemblea Legistlative do Paraná no dia 15 de Feberero de 1876 pelo Presidente da Provincia da o excellentissimo senhor doutor Adolpho Lins*, pp. 82–3; Paraná, *Relatório apresentado a Assemblea Legistlative do Paraná, publicado como anexo dos relatórios do 23 de Feberero de 1878 e do 9 de Abril 1878 pelo Presidente da Provincia da o excellentissimo senhor doutor Bento de Oliveira Júnior* (Curitiba, 1878), p. 53.
42 *The Anglo-Brazilian Times*, 22 April 1870.
43 *The Anglo-Brazilian Times*, 7 July 1874.

44 The Anglo-Brazilian Times, 23 October 1874.
45 The Anglo-Brazilian Times, 23 March 1875.
46 Ibid.
47 Bocaiúva, 'Uma questão pessoal', p. 426.
48 Lesser, Negotiating National Identity, pp. 22–4. Bocaiúva and his allies (amongst whom was the government's former Inspector of Colonies, Ignácio da Cunha Galvão) persuaded the government to embark on a program of encouraging Chinese immigration. In complete contrast to Luiz Werneck's earlier opinions, the Chinese, to Bocaiúva, were calm, well-mannered and hard working, characteristics that he thought would make them capable of enduring the often-harsh conditions of life and labour on a Brazilian coffee plantation. Although Bocaiúva conceded that the Chinese also came with negative qualities, these were not considered overly important as their supposed strong attachment to the land of their ancestors meant that they would not want to remain in Brazil beyond the period stipulated in their contracts. In July 1870 a ten-year plan to promote Chinese immigration was enacted, though nothing was to come of this, in part due to British and Chinese actions to clamp down on the near slave-like international "coolie trade".
49 José C. Moya, Cousins and Strangers: Spanish Immigrants in Buenos Aires, 1850–1930 (Berkeley: University of California Press, 1998), pp. 72–5.
50 See Platt 'British agricultural colonization in Latin America, Part 1', 3–38 and Platt, 'British agricultural colonization in Latin America, Part 2', 23–42. One such scheme that attracted poor Londoners – including some English families, but mainly Jewish eastern European immigrants – in the 1870s involved Paraguay as a destination. See Juan Carlos Herken-Krauer, 'La inmigración en la Paraguay de posguerra: el caso de los "Lincolnshire Farmers" (1870–1873)', Revista Paraguaya de Sociología 18/52 (1981), 33–107; Harris Gaylord Warren, 'The "Lincolnshire Farmers" in Paraguay: an abortive emigration scheme of 1872–1873", The Americas 21 (1965), 243–62.
51 The Board claimed, however, that it was powerless to intervene beyond issuing a public warning to would-be emigrants not to proceed and by making regular appeals to the Brazilian government either to transfer British nationals to better land or return them to England without charge; Murdoch to Standford, 17 December 1869, NA/PRO FO128/93; Rogers (Colonial Office) to Foreign Office, 22 March 1870, NA/PRO CO386/119.
52 Browne, 'Government immigration policy in imperial Brazil, 1822–1870', p. 351.

Epilogue

1 Statement signed by George Stewart Boddy on behalf of British residents of Cêrro Azul (formerly Assunguy) to Acting Consul Hampshire (Santos), 9 November 1887, NA/PRO FO128/155.
2 Tigar (Assunguy) to Drummond (Rio de Janeiro), 9 January 1876, NA/PRO FO128/108/289; The Anglo-Brazilian Times, 9 April 1878.
3 This is not the case of many descendants of British immigrants who live in Curitiba, many of whom keenly pursue genealogical interests. Although Portuguese, Polish, Ukrainian, German, Italian and Japanese surnames – the legacies of the far more successful post-1875 currents of immigration – are far more common in Paraná, English and Irish surnames are probably less unusual there than elsewhere in Brazil. Bond, Brine, Boddy, Chamberlain, Cooper, Pugsley, Samways and Tamplin are just a few of the surnames linked with Colônia Assunguy that one comes across today in Curitiba, with these descendants of British immigrants found at every level of society within the state of Paraná.

Notes

4 Impressions of Cêrro Azul are based on visits by the author in 1997 and 1999. Additional information has been kindly supplied in 2004 by Mauro Fortes Carneiro of Curitiba.
5 John Mitchell (Cananéia) to Gastrell (Rio de Janeiro), 15 January 1881, NA/PRO FO128/116.
6 James Davies (Cananéia) to Corbett (Rio de Janeiro), 10 May 1884, NA/PRO FO128/138.
7 Davies (Cananéia) to Corbett (Rio de Janeiro), 10 May 1884, NA/PRO FO128/138; Cowper (Santos) to British Legation (Rio de Janeiro), 25 April 1887, NA/PRO FO128/150.
8 May (HMS *Rifleman*, Rio de Janeiro), 2 October 1888, NA/PRO FO128/157; Bridget Davies (Cananéia), 24 December 1886, NA/PRO FO128/146.
9 May (HMS *Rifleman*, Rio de Janeiro), 2 October 1888, NA/PRO FO128/157.
10 F.D. Lilly, surgeon (HMS *Rifleman*, Santos), 29 September 1888, NA/PRO FO128/157; May (HMS *Rifleman*, Rio de Janeiro), 2 October 1888, NA/PRO FO128/157.
11 Impressions of the Ex-Colônia are based on a visit by the author in December 2002.

Appendix 2

1 Bridget Davies (Cananéia) to British Minister (Rio de Janeiro), 25 September 1898, NA/PRO FO128/240.

Picture Credits

Page 55, Black Country townscape: William E. Littlewood, *Down in Dingyshire; or, Sketches of Life in the Black Country* (London: Seeley, Jackson & Halliday, 1873); British Library (4192.b.35).

Page 71, "Dr" Barziller Cottle in 1867: University of Nevada, Reno Library, Special Collections (1023/I).

Page 95, Field hands taking a refreshment break: *The Illustrated London News*, 3 July 1875.

Page 104, Advertising for Brazil: *The Labourers' Union Chronicle*, 18 January 1873; Warwickshire County Record Office.

Page 107, Agricultural labourers meeting in Wellesbourne, Warwickshire, 1872: *The Illustrated London News*, 27 April 1872.

Page 115, The emigrant's last sight of home: Richard Redgrave, 1858; © Tate, 2005.

Page 129, Fazenda Montebello, Espírito Santo: Albert Richard Dietze, c. 1875; Coleção Dona Thereza Christina Maria, Fundação Biblioteca Nacional, Rio de Janeiro.

Page 154, A panic upon a bridge: Thomas P. Bigg-Wither, *Pioneering in South Brazil*, vol. 2 (London: John Murray, 1878); British Library (2374.b.17).

Page 160, The first stage of clearing the forest for settlement: Bigg-Wither, *Pioneering in South Brazil*, vol. 2; British Library (2374.b.17).

Page 163, 'The Grange': the Tamplin family's farm: Caroline Tamplin, c. 1870s; Tigar family (Vancouver).

Page 168, The Turvo Valley schoolhouse: Caroline Tamplin, c. 1870s; Tigar family (Vancouver).

Page 209, Wedding gathering, Cêrro Azul, c. 1927; Margarida Chamberlain (Curitiba).

Page 210, Three generations of the Chamberlain family, Cêrro Azul, 1998; Oliver Marshall.

Page 211, Luiza da Conceição Blane and Ernesto Fitz, Turvo (Cêrro Azul), 1998; Oliver Marshall.

Page 215, João and David Davies, Ex-Côlonia (Cananéia), 2002; Oliver Marshall.

Bibliography

Primary Sources

Archival collections

Arquivo Histórico e Diplomático, Palacio Itamaraty, Rio de Janeiro
Arquivo Público do Paraná (APP), Curitiba, Paraná
Birmingham Archdiocesan Archives, St Chad's Cathedral, Birmingham
British Library (Department of Manuscripts), London
Connelly-Orgill letters (Helen Jean Gent collection), Pittsburgh, Pennsylvania
Grenfell Papers (Corrêa do Lago collection), São Paulo
Family Records Centre, London
National Archives (NA/PRO), London
    Colonial Office (CO)
    Foreign Office (FO)
    Home Office (HO)
National Archives (NA/US), Washington, DC
University of Nevada Library Special Collections, Reno, Nevada
Perkins Library, Duke University, Durham, North Carolina
Caroline Tamplin diaries (Tamplin family collection), Curitiba, Paraná
Albert Tigar manuscript (Tigar family collection), Vancouver, British Columbia
United States, Department of State, National Archives (NA/US), Washington, DC

Published official reports

Brazil, *Relatório do Ministério e Secretario de Estado dos Negocios da Agricultura, Commercio e Obras Publicas* (Rio de Janeiro, 1868).

Brazil, *Relatório do Ministério e Secretario de Estado dos Negocios da Agricultura, Commercio e Obras Publicas* (Rio de Janeiro, 1869).

Brazil, *Relatório do Ministério e Secretario de Estado dos Negocios da Agricultura, Commercio e Obras Publicas* (Rio de Janeiro, 1870).

Brazil, *Relatório do Ministério e Secretario de Estado dos Negocios da Agricultura, Commercio e Obras Publicas* (Rio de Janeiro, 1872).

Brazil, *Relatório do Ministério e Secretario de Estado dos Negocios da Agricultura, Commercio e Obras Publicas* (Rio de Janeiro, 1875).

Great Britain, *Reports Respecting the Condition of British Emigrants in Brazil* (London: Parliamentary Paper, vol. 75, 1873).

Great Britain, *Report Respecting the Condition of British Emigrants in Brazil* (London: Parliamentary Paper, vol. 76, no. 3, 1874).

Great Britain, *Emigration to Brazil. Report on the Colony of Assunguy* (London: Parliamentary Paper, vol. 82, 1875).

Great Britain, *Emigration to Brazil. Report on the Colony of Cananea* (London: Parliamentary Paper, vol. 82, 1875).

Great Britain, *Thirty-third Annual General Report of the Emigration Commissioners* (London: Parliamentary Papers, vol. 18, 1875).

Paraná, *Relatório apresentado a Assembléia Legistlativa do Paraná, 1860, pelo Presidente da Provincia da o senhor José Francisco Cardoso* (Curitiba, 1860).

Paraná, *Relatório apresentado a Assembléia Legistlativa do Paraná no dia 15 de Feberero de 1872 pelo Presidente da Provincia da o excellentissimo senhor Oliveira Lisboa* (Curitiba, 1872).

Paraná, *Relatório apresentado a Assembléia Legistlativa do Paraná no dia 15 de Feberero de 1876 pelo Presidente da Provincia da o excellentissimo senhor doutor Adolpho Lins* (Curitiba, 1876).

Paraná, *Relatório apresentado a Assemblea Legistlativa do Paraná, publicado como anexo dos relatórios do 23 de Feberero de 1878 e do 9 de Abril 1878 pelo Presidente da Provincia da o excellentissimo senhor doutor Bento de Oliveira Júnior* (Curitiba, 1878).

Rio Grande do Sul, *Relatório do Vice-Presidente da Provincia de S. Pedro do Rio Grande do Sul, Luiz Aloes Leite de Oliveira Bello, na Abertura da Assembléa Legislativa Provincial em 01 de Outubro de 1852* (Porto Alegre, 1852).

Santa Catarina, *Relatórios apresentados á Assembléa Legislativa Provincial de Santa Catarina, na sua sessão ordinaria, e ao Vice-Presidente Commendador Francisco José de Oliveira, no anno de 1868* (Rio de Janeiro, 1868).

Newspapers and periodicals

The Anglo-Brazilian Times (Rio de Janeiro)

The Banbury Guardian (Banbury)

The Bee-Hive (London)

Birmingham Daily Post (Birmingham)

The Brazil and River Plate Mail (London)

The Cork Examiner (Cork)

Daily News (London)

The Daily Telegraph (London)

Bibliography

O Dezenove de Dezembro (Curitiba, Paraná)

The Dudley Herald and Wednesbury Borough News (Dudley)

The Gloucester Journal (Gloucester)

Gloucester Mercury (Gloucester)

The Illustrated London News (London)

The Irish Catholic Chronicle (Dublin)

Die Kolonie Zeitung (Joinville, Santa Catarina)

The Labourers' Union Chronicle (Leamington)

The Liverpool Journal of Commerce (Liverpool)

The Morning Post (London)

The Rev. G. Montgomery's Register (Wednesbury)

The Royal Leamington Spa Courier and Warwickshire Standard (Leamington)

The Royal Leamington, Warwickshire, & Centre of England Chronicle (Leamington)

The Rugby Advertiser (Rugby)

South American Missionary Magazine (London)

The Times (London)

The Universal News (London)

The Wednesbury and West Bromwich Advertiser (Wednesbury)

The Weekly Register & Catholic Standard (London)

The Wilts and Gloucestershire Standard (Cirencester)

Other published primary sources

Arch, Joseph, *Joseph Arch: The Story of His Life, Told by Himself* (London: Hutchinson, 1898).

Bigg-Wither, Thomas P., *Pioneering in South Brazil*, 2 vols, (London: John Murray, 1878).

Bocaiúva, Quintino, 'Uma questão pessoal: carta a Ilustrada Redação de "*A Província de São Paulo*" (1877)', in Eduardo Silva, ed., *Idéias políticas de Quintino Bocaiúva* (Rio de Janeiro: Fundação Casa de Rui Barbosa, 1986), pp. 425–34.

Buckle, Henry Thomas, *History of Civilization in England*, vol. I, (London: John W. Parker, 1857).

Burritt, Elihu, *A Walk from London to Land's End and Back* (London: Sampson, Low, Son, and Marston, 1865).

_____, *Walks in the Black Country and its Green Border-land* (London: Sampson, Low, Son, and Marston, 1868).

Burton, Isabel (with W.H. Wilkins), *The Romance of Isabel, Lady Burton: The Story of Her Life* (London: Hutchinson, 1897).

Burton, Richard F., *Explorations of the Highlands of Brazil*, vol. 1, (London: Tinsley Brothers, 1869).

Carvalho, Augusto de, *O Brazil: colonisação e emigração* (Porto: Imprensa Portugueza, 1876).

Codman, John, *Ten Months in Brazil; with Notes on the Paraguayan War* (Edinburgh: R. Grant & Son, 1870).

Duffy, Patrick J., ed., *To and from Ireland: Planned Migration Schemes, c. 1600–2000* (Dublin: Geography Publications, 2004).

Dunlop, Charles, *Brazil as a Field for Emigration: its Geography, Climate, Agricultural Capabilities, and the Facilities Afforded for Permanent Settlement* (London: Bates, Hendy & Co., 1866).

Emigration to Brazil (New York: S. Hallet, 1866).

Formby, Henry, *A Voice from the Grave: Being the Funeral Discourse of the Rev. George Montgomery* (London: Burns, Oates, and Company, 1871).

Hadfield, William, *Brazil and the River Plate, 1870–76* (Sutton, Surrey: W.R. Church, 1877).

Handbook for Emigrants to Brazil. Containing a collection [.....] Brazilian Legislation that Most Particulary [sic] Interest Those Strangers Who Will Make Their Residence in Brazil, etc. (Rio de Janeiro: E. & H. Laemmert, 1865).

Hardy, Thomas, 'The Dorset farm labourer', *Longman's Magazine* 2 (May–Oct 1883), 252–69.

_____, *Tess of the D'Urbervilles* (London: Macmillan, 1963; first published 1891).

Heath, Francis G., *The English Peasantry* (London: Frederick Warne and Co., 1874).

_____, *British Rural Life and Labour* (London: P.S. King & Son, 1911).

Heinrick, Hugh, *A Survey of the Irish in England (1872)*, edited by Alan O'Day (London: The Hambledon Press, 1990).

"Jacaré Assu", pseud., *Brazilian Colonization from a European Point of View* (London: Edward Stanford, 1873).

Kidder, Daniel P. and James C. Fletcher, *Brazil and the Brazilians: Portrayed in Historical and Descriptive Sketches* (Philadelphia, PA: Childs & Peterson and London: Bates, Hendy & Co., 1857).

_____, *Brazil and the Brazilians* (Boston, MA: Little, Brown and Co., 1866).

Lee, John Irwin, *Three Hundred British Immigrants, Many of Whom are Dying From Starvation in the Colony of Assunguy, Paraná, Brazil, South America* (London: William Macintosh & Co., 1875).

Littlewood, William E., *Down in Dingyshire; or, Sketches of Life in the Black Country* (London: Seeley, Jackson & Halliday, 1873).

Mansfield, Charles Blachford, *Paraguay, Brazil, and the Plate* (Cambridge: Macmillan & Co., 1856).

Mulhall, Marian, *Between the Amazon and Andes, or Ten Years of a Lady's Travels in the Pampas, Gran Chaco, Paraguay, and Matto Grosso* (London: Edward Stanford, 1881).

Mulhall, Michael G., *Rio Grande do Sul, and its German Colonies* (London: Longman, Green and Co., 1873).

Murray, John Hale, *Travels in Uruguay, South America* (London: Longmans & Co., 1871).

Olmsted, Frederick Law, *Walks and Talks of an American Farmer in England* (Ann Arbor: Michigan University Press, 1969).

Scully, William, *Brazil; Its Provinces and Chief Cities; The Manners and Customs of the*

People; Agricultural, Commercial and Other Statistics, Taken From the Latest Official Documents; with a Variety of Useful and Entertaining Knowledge, Both for the Merchant and the Emigrant (London: Murray & Co., 1866).

Smith, Herbert H., *Brazil: the Amazons and the Coast* (New York: Charles Scribner's Sons, 1879).

Sociedade Internacional de Imigração, *Relatorio Annual da Directoria*, no. 1, (Rio de Janeiro: Typographia Imperial e Constitucional de J. Villeneuve & Cia., 1867).

Society for Promoting Christian Knowledge, *Birmingham and the Black Country* (London: Society for Promoting Christian Knowledge, 1864).

Southey, Robert, *History of Brazil* (London: Longman, Durst, Rees, Orme, and Brown, 1824).

Sturz, Johann Jacob, *Die deutsche Auswanderung und die Verschleppung deutscher Auswanderer* (Berlin: Kortkampf, 1868).

Ursel, Charles d', *Sud-Amérique: séjours et voyages au Brésil, à la Plata, au Chili, en Bolivie et au Pérou* (Paris: E. Plon, 1879).

Werneck, Luiz Peixoto de Lacerda, *Idéias sobre colonização precedidas de una succinta exposição dos principios geraes que regem a população* (Rio de Janeiro: Eduardo e Henrique Laemmert, 1855).

Whitehead, Charles, *Agricultural Labourers* (London: Longmans, Green, Reader, and Dyer, 1870).

Secondary sources

Abbott, G. J., 'The emigration to Valparaiso in 1843', *Labour History* (Canberra) 19 (1970), 1–16.

Almeida, Antônio Paulino de, *Memória histórica sôbre Cananéia*, vol. 1, (São Paulo: Universidade de São Paulo, Faculdade de Filosofia, Ciêncas e Letras, 1963).

Araujo Neto, Miguel Alexandre de, 'British journalism in late 19th Century Brazil: the case of the Anglo-Brazilian Times (1865–1884)', unpublished MA thesis, Institute of Latin American Studies, University of London, 1992.

_____, 'Imagery and arguments pertaining to the issue of free immigration in the Anglo-Irish press in Rio de Janeiro', *ABEI Journal: The Brazilian Journal of Irish Studies* (São Paulo) 5 (2003), 111–27.

Arnstein, Walter L., 'The Murphy riots: a Victorian dilemma', *Victorian Studies* 19 (1975), 51–71.

Arnold, Rollo, *The Farthest Promised Land: English Villagers, New Zealand Immigrants of the 1870s* (Wellington: Victoria University Press, 1981).

Baines, Dudley, *Migration in a Mature Economy: Emigration and Internal Migration in England and Wales, 1861–1900* (Cambridge: Cambridge University Press, 1985).

Barnsby, George J., *Social Conditions in the Black Country, 1800–1900* (Wolverhampton: Integrated Publishing Services, 1980).

Basto, Fernando L.B., *Ex-Combatentes Irlandeses em Taperoa* (Rio de Janeiro: Editôra Vozes, 1971).

Belchem, John, 'Class, creed and country: the Irish middle class in Victorian Liverpool,' in Roger Swift and Sheridan Gilley, eds., *The Irish in Victorian Britain: The Local Dimension* (Dublin: Four Courts Press, 1999), pp. 190–211.

Bethell, Leslie, 'The independence of Brazil', in Leslie Bethell, ed., *Brazil: Empire and Republic, 1822–1930* (Cambridge: Cambridge University Press, 1989), pp. 3–42.

_____, 'Britain and Latin America in historical perspective', in Victor Bulmer-Thomas, ed., *Britain and Latin America: A Changing Relationship* (Cambridge: Cambridge University Press, 1989), pp. 1–24.

Birch, Brian, 'Popularizing the Plains: news of Kansas in England, 1860–1880', *Kansas History* 10/4 (1987/88), 262–74.

Bromfield, J.F., *St.Mary's Centenary Souvenir* (Wednesbury, 1950).

Browne, George P., 'Government immigration policy in imperial Brazil, 1822–1870' unpublished PhD thesis, The Catholic University of America, 1972.

Carey, John, 'Did the Irish come from Spain? The legend of the Milesians', *History Ireland* 9/3 (2001), 8–11.

Carneiro, José Fernando, *Imigração e colonização no Brasil* (Rio de Janeiro, 1949).

Chatwin, Bruce, *In Patagonia* (London: Jonathan Cape, 1977).

Conrad, Robert, 'The planter class and the debate over Chinese immigration to Brazil, 1850–1893', *International Migration Review* 9/1 (1975), 41–55.

Costa, Emilia Viotti da, *The Brazilian Empire: Myths and Histories* (Chicago: University of Chicago Press, 1985).

Davis, Graham, *Land! Irish Pioneers in Mexican and Revolutionary Texas* (College Station: Texas A&M University Press, 2002).

Diner, Hasia R., '"The most Irish city in the Union": the era of the Great Migration, 1844–1877', in Ronald H. Bayor and Timothy J. Meagher, eds., *The New York Irish* (Baltimore, MD: The Johns Hopkins University Press, 1996), pp. 87–106.

Donnelly, Jr., James S., *The Great Irish Potato Famine* (Stroud: Sutton Publishing, 2001).

Donovan, Bill M., 'The politics of immigration to eighteenth-century Brazil: Azorean migrants to Santa Catarina', *Itinerario* 16/1 (1992), 35–53.

Dunbabin, J.P.D., 'The "Revolt of the Field": the agricultural labourers' movement in the 1870s', *Past and Present* 26 (1963), 68–98.

Eakin, Marshall C., *British Enterprise in Brazil: The St. John d'el Rey Mining Company and the Morro Velho Gold Mine, 1830–1960* (Durham, NC: Duke University Press, 1989).

Ede, John F., *History of Wednesbury* (Wednesbury: Wednesbury Corporation, 1962).

Emrich, Oswaldo S., *Histórico da Igreja Presbiteriana de Curitiba: primeiro centenario, 1888–1988* (Curitiba: Histórico da Igreja Presbiteriana de Curitiba, 1988)

Erickson, Charlotte, *Invisible Immigrants: The Adaptation of English and Scottish Immigrants in Nineteenth-Century America* (Ithaca, NY: Cornell University Press, 1990).

_____, *Leaving England: Essays on British Emigration in the Nineteeenth Century* (Ithaca, NY: Cornell University Press, 1994).

Fairburn, Miles, *The Ideal Society and its Enemies: The Foundations of Modern New Zealand Society, 1850–1900* (Auckland: Auckland University Press, 1989).

Ficker, Carlos, *Colonos de Joinville na guerra do Paraguai* (Blumenau: Blumenau em Cadernos, 1966).

_____, *Charles van Lede e a colonização belga em Santa Catarina. Subsídios para a história de colonização de Ilhota, no rio Itajaí-Açu, pela 'Compagne belge-brésilienne de colonisation'* (Blumenau: Blumenau em Cadernos, 1972).

Fitzpatrick, David, *Oceans of Consolation: Personal Accounts of Irish Migration to Australia* (Cork: Cork University Press, 1994).

Freyre, Gilberto, *Ingleses* (Rio de Janeiro: José Olympio, 1942).

_____, *Ingleses no Brasil: aspectos da influencia Britânica sobre a vida, a paisagem e a cultura do Brasil* (Rio de Janeiro: José Olympio, 1948).

Gillow, Joseph, *A Literary and Biographical History, or Bibliographical Dictionary of the English Catholic*, vol. 5, (London: Burns & Oates, 1902).

Goldman, Frank P., *Os pioneiros americanos no Brasil: educadores, sacerdotes, covos e reis* (São Paulo: Livraria Pioneira Editora, 1972).

Gordon, Michael A., *The Orange Riots: Irish Political Violence in New York City, 1870 and 1871* (Ithaca, NY: Cornell University Press, 1993).

Graham, Richard, *Britain and the Onset of Modernization in Brazil, 1850–1914* (Cambridge: Cambridge University Press, 1968).

Green, Frederick E., *A History of the English Agricultural Labourer, 1870–1920* (London: P.S. King & Son, 1920).

Griggs, William Clark, *The Elusive Eden: Frank McMullan's Confederate Colony in Brazil* (Austin: University of Texas Press, 1987).

Groves, Reg, *Sharpen the Sickle! The History of the Farm Workers' Union* (London: The Merlin Press, 1981).

Guenther, Louise, *British Merchants in Nineteenth-Century Brazil: Business, Culture, and Identity in Bahia, 1808–50* (Oxford: Centre for Brazilian Studies, University of Oxford, 2004).

Hackwood, Frederick William, *Religious Wednesbury: Its Creeds, Churches & Chapels* (Dudley: The Herald Press, 1900).

Harris, Ruth-Ann, *The Nearest Place That Wasn't Ireland: Early Nineteenth-Century Irish Labor Migration* (Ames: Iowa State University Press, 1994).

Harnsberger, John L. and Robert P. Wilkins, 'New Yeovil, Minnesota: a Northern Pacific colony in 1873', *Arizona and the West* 12/1 (1970), 5–22.

Hemming, John, *Amazon Frontier: The Defeat of the Brazilian Indians* (London: Macmillan, 1987).

Hering, Maria Luiza Renaux, *Colonização e indústria no Vale do Itajaí: o modelo catarinense de desenvolvimento* (Blumenau: Editora da Fundação Universidade de Blumenau, 1987).

Herken-Krauer, Juan Carlos, 'La inmigración en la Paraguay de posguerra: el caso de los "Lincolnshire Farmers" (1870–1873)', *Revista Paraguaya de Sociología* 18/52 (1981), 33–107.

Hill, Lawrence F., *Diplomatic Relations Between the United States and Brazil* (Durham, NC: Duke University Press, 1932).

Hobsbawm, Eric J. and George Rudé. *Captain Swing* (London: Lawrence & Wishart, 1969).

Hodgkins, John R., *Over the Hills to Glory: Radicalism in Banburyshire, 1832–1945* (Southend: Clifton Press, 1978).

Horn, Pamela, 'Gloucestershire and the Brazilian emigration movement, 1872–73', *Transactions of the Bristol and Gloucestershire Archaeological Society* LXXXIX (1970), 167–74.

_____, 'Agricultural trade unionism and emigration, 1872–1881', *The Historical Journal* 15 (1972), 87–102.

_____, *Labouring Life in the Victorian Countryside* (Dublin: Gill and Macmillan, 1976).

Jarnagin, Laura, 'Relocating family and capital within the nineteenth-century Atlantic World economy: the Brazilian connection', in Cyrus B. Dawsey and James M. Dawsey, eds., *The Confederados: Old South Immigrants in Brazil* (Tuscaloosa: University of Alabama Press, 1995), pp. 66–83.

Jones, Maldwyn A., 'The background to emigration from Great Britain in the nineteenth century', *Perspectives in American History* 7 (1973), 3–92.

Kellenbenz, Hermann and Jürgen Schneider, 'A imagem do Brasil na Alemanha do século XIX: impressões e estereótipos da independencia ao fim do monarquia', *Estudios Latinoamericanos* (Warsaw) 6/2 (1980), 78–81.

Korol, Juan Carlos and Hilda Sábato. *Cómo fue la inmigración irlandesa en la Argentina* (Buenos Aires: Plus Ultra, 1981).

Kristjanson, Wilhelm, *The Icelandic People in Manitoba* (Winnipeg: Wallingford Press, 1963).

Lamb, Roberto Edgar, *Uma jornada civilizadora: imigração, conflito social e segurança pública na Provincia do Paraná, 1867–1882* (Curitiba: Aos Quatro Ventos, 1997).

Lauth, Aloisius Carlos, *A Colônia Príncipe Dom Pedro: um caso de política no Brasil Império* (Brusque: Privately published, 1987).

Lemos, Juvêncio Saldanha, *Os mercenários do Imperador: a primeira corrente imigratória alemã no Brasil* (Porto Alegre: Palmarinca, 1993).

Lesser, Jeffrey, 'Neither slave nor free, neither black nor white: the Chinese in early nineteenth century Brazil', *Estudios Interdisciplinarios de América Latina y el Caribe* 5/2 (1994), 23–34.

_____, *Negotiating National Identity: Immigrants, Minorities, and the Struggle for Ethnicity in Brazil* (Durham, NC: Duke University Press, 1999).

Macaulay, Neill, *Dom Pedro: The Struggle for Liberty in Brazil and Portugal, 1798–1834* (Durham, NC: Duke University Press, 1986).

Macedo, Francisco Riopardense de, *Ingleses no Rio Grande do Sul* (Porto Alegre: Edições a Nação, 1975).

Machado-Curry, Jonathan, 'Indispensable aliens: the influence of engineering migrants in mid-nineteenth century Cuba', unpublished PhD thesis, London Metropolitan University, 2003.

McKay, Ernest A., *The Civil War and New York City* (Syracuse, NY: Syracuse University Press, 1990).

Bibliography

McKenna, Patrick, 'Irish society in nineteenth century Argentina', in Oliver Marshall, ed., *English-Speaking Communities in Latin America* (Basingstoke: Macmillan, 2000), pp. 81–103.

Malchow, Howard L., 'Trade unions and emigration in late Victorian England: a national lobby for state aid', *The Journal of British Studies* 15/2 (1976), 92–116.

Martins, Wilson, *Um Brasil diferente: ensaio sobre fenôminos da aculturação no Parané*, 2. ed., (São Paulo: T.A. Queiroz, 1989).

Mello, José Antônio Gonsalves de, *Ingleses em Pernambuco: História do Cemitério Britânico do Recife e da participação de ingleses e outros estrangeiros na vida e na cultura de Pernambuco, no período de 1813 a 1909* (Recife: Instituto Arqueológico, Histórico e Geográfico Pernambucano, 1972).

Mendonça, Antonio Gouvêa, 'Protestantes na diáspora', in Martin N. Dreher, ed., *Imigrações e história da igreja no Brasil* (São Paulo: Editora Santuário, 1993), pp. 132–57.

Millburn, John R. and Keith Jarrott, *The Aylesbury Agitator: Edward Richardson, the Labourers' Friend and Queensland Agent, 1849–1878* (Aylesbury: Buckinghamshire County Library, 1988).

Miller, Kerby A., *Emigrants and Exiles: Ireland and the Irish Exodus to North America* (New York: Oxford University Press, 1985).

Miller, Rory, *Britain and Latin America in the Nineteenth and Twentieth Centuries* (London: Longman, 1993).

Mingay, Gordon E., *Rural Life in Victorian England* (Stroud: Alan Sutton, 1990).

_____, ed., *The Victorian Countryside*, 2 vols. (London: Routledge & Kegan Paul, 1981).

Moog, Víanna, *Bandeirantes and Pioneers* (New York: George Braziller, 1964).

Moran, Gerard, '"In search of the promised land": the Connemara colonization scheme to Minnesota, 1880', *Éire-Ireland* 31/3–4 (1996), 130–49.

Moya, José C., *Cousins and Strangers: Spanish Immigrants in Buenos Aires, 1850–1930* (Berkeley: University of California Press, 1998).

Murphy, Hilary, 'When Wexford farmers emigrated to Brazil', *Journal of the Irish Family History Society* 2 (1986), 41–3.

Murray, Edmundo, 'How the Irish became *gauchos ingleses*: diasporic models in Irish–Argentine literature', unpublished Mémoire de Licence, L'Université de Geneve, 2003.

_____, *Devenir irlandés: narrativas íntimas de la emigración irlandesa a la Argentina (1844–1912)* (Buenos Aires: Eudeba, 2004).

Nicoulin, Martin, *A Gênese de Nova Friburgo: emigração e colonização suíça no Brasil, 1817–1827* (Rio de Janeiro: Fundação Biblioteca Nacional, 1996).

Nunes. Rosana Barbosa, 'Portuguese migration to Rio de Janeiro, 1822–1850', *The Americas* 57/1 (2000), 37–61.

O'Donnell, Edward, '"The scattered debris of the Irish nation": the famine Irish and New York City', in E. Margaret Crawford, ed., *The Hungry Stream: Essays on Emigration and Famine*, (Omagh: Ulster-American Folk Park, Centre for Emigration Studies and Belfast: Queen's University of Belfast, Institute of Irish Studies, 1997), pp. 49–60.

O'Neill, Peter, ed., *Links between Brazil & Ireland 1998/99 Survey* (Rio de Janeiro: Peter O'Neill, 1999).

O'Sullivan, Patrick, ed., *St Mary's Wednesbury 1850–2000* (Wednesbury: St Mary's Church, 2000).

Park, Kyeyoung, '"I am floating in the air": creation of a Korean transnational space among Korean-Latino American remigrants', *Positions: East Asia Cultures Critique* 7/3 (1999), 667–95.

Platt, D.C.M., 'British agricultural colonization in Latin America, Part I' *Inter-American Economic Affairs* 18/3 (1964), 3–38.

_____, 'British agricultural colonization in Latin America, Part 2' *Inter-American Economic Affairs* 19/1 (1965), 23–42.

Queiroz, M.I.P., 'Messiahs in Brazil.' *Past and Present* 31 (1965), 62–86.

Redouane, Joëlle Annie, 'The Irish in Algeria, 1830–1930', *Éire-Ireland* 21/4 (1986), 3–10.

Rees, Jim, *Surplus People: The Fitzwilliam Clearances, 1847–1856* (Cork: The Collins Press, 2000).

Reid, George, *A South American Adventure: Letters from George Reid, 1867–1870, Selected by Valerie Boyle* (London: Valerie Boyle, 1999).

Rocha, Gilda, *Imigração estrangeira no Espírito Santo, 1847–1896* (Vitória: Privately published, 2000).

Rocha-Trindade, Maria Beatriz, 'Portuguese migration to Brazil in the nineteenth and twentieth centuries: an example of international cultural exchanges', in David Higgs, ed., *Portuguese Migration in Global Perspective* (Toronto: Multicultural History Society of Ontario, 1990), pp. 29–41.

Rozek, Barbara J., *Come to Texas: Enticing Immigrants, 1865–1915* (College Station, TX: Texas A&M University Press, 2003).

Salzman, L. F., ed., *The Victoria History of the County of Warwick*, vol. VI, (London: Oxford University Press, 1951).

Santos, Sílvio Coelho dos, *Indios e brancos no sul do Brasil: a dramática experiência dos Xokleng* (Porto Alegre: Editora Movimento, 1987).

Scotland, Nigel, *Agricultural Trade Unionism in Gloucestershire, 1872–1950* (Cheltenham: The Centre for the Study of Religion, Cheltenham and Gloucester College of Higher Education, 1991).

Seyferth, Giralda, *Colonização e conflito: estudo sobre "motins" e "desordens" numa região colonial de Santa Catarina no século XIX* (Rio de Janeiro: Museu Nacional – UFRJ, Programa de Pos-graduação em Antropologia Social, 1988).

_____, *A colonização alemão no Vale do Itajaí-Mirim: um estudo de desenvolvimento econômico*, 2. ed., (Porto Alegre: Editora Movimento, 1999).

Silva, Eduardo, *Barões e escravidão* (Rio de Janeiro: Nova Fronteira, 1984).

Simpson, Tony, *The Immigrants: The Great Migration from Britain to New Zealand, 1830–1890* (Auckland: Godwit Publishing, 1997).

Skidmore, Thomas E., *Black into White: Race and Nationality in Brazilian Thought* (Durham, NC: Duke University Press, 1993).

Souter, Gavin, *A Peculiar People: The Australians in Paraguay* (Sydney: Sydney University Press, 1981).

Spitzer, Leo, *Hotel Bolivia: The Culture of Memory in a Refuge from Nazism* (New York: Hill and Wang, 1998).

Sutherland, Daniel E., *The Confederate Carpetbaggers* (Baton Rouge: Louisiana State University Press, 1988).

Bibliography

Swift, Roger, 'The historiography of the Irish in nineteenth-century Britain', in Patrick O'Sullivan, ed., *The Irish in the New Communities* (Leicester: Leicester University Press, 1992), pp. 99–127.

Taylor, Philip A.M., *The Distant Magnet: European Emigration to the U.S.A.* (New York: Harper & Row, 1971).

_____, 'The Distant Magnet after twenty-five years: an essay in self-criticism', *Journal of American Studies* 31/2 (1997), 295–312.

Thistlethwaite, Frank, 'Migration from Europe overseas in the nineteenth and twentieth centuries', in *Rapports V Histoire Contemporaine, XIe Congres International des Sciences Historiques, Stockholm* (Göteborg: Almqvist & Wiksell, 1960), pp. 32–60.

Thomas, Lewis H., 'Welsh settlement in Saskatchewan.' *Western Historical Quarterly* 4/4 (1973), 435–49.

Thorsteinsson, Thorsteinn, *Aefrintyrith. Fra Isandi til Brasiliu. Fyrstu folksflutningarnir fra Northurlandi* [*The Adventure. From Iceland to Brazil*] (Reykjavik: Sigurgeir Frithriksson, 1937–8).

Towell, Larry, *The Mennonites: A Biographical Sketch* (London: Phaidon, 2000).

Vale, Brian, 'British sailors and the Brazilian navy, 1822–1850', *The Mariner's Mirror* 80/3 (1994), 312–25.

Wachowicz, Rui, 'O comércio da madeira e a atuação da Brazil Railway no sul do Brasil.' *Boletim do Instituto Histórico, Geografico e Etnográfico Paranaense* 42 (1984), pp. 41–78.

Walker, Mack, *Germany and the Emigration, 1816–1885* (Cambridge, MA: Harvard University Press, 1964).

Ward, Shelagh Mary, 'Pennies of the poor: Catholics and poverty in Bradford, 1860–1914', unpublished MA thesis, Victorian Studies, Trinity and All Saints College, University of Leeds, 2000.

Warren, Harris Gaylord, 'The "Lincolnshire Farmers" in Paraguay: an abortive emigration scheme of 1872–1873", *The Americas* 21 (1965), 243–62.

Weaver, Blanche Henry Clark, 'Confederate emigration to Brazil', *The Journal of Southern History* 27/1 (1961), 33–53.

Wey, Claude, 'Luxembourgers in Latin America and the permanent threat of failure: "return migration" in the social context of a European micro-society', *AEMI Journal* 1 (2003), 1–11.

Williams, Glyn, *The Desert and the Dream: A Study of Welsh Colonization in Chubut, 1865–1915* (Cardiff: University of Wales Press, 1975).

_____, 'La imagen sobre América Latina en Gales durante los siglos XIX y XX', *Estudios Latinoamericanos* (Warsaw) 6/2 (1980), 370–7.

Williams, R. Bryn, *Y Wladfa* (Cardiff: University of Wales Press, 1962).

Wtulich, Josephine, ed., *Writing Home: Immigrants in Brazil and the United States, 1890–1891* (Boulder, CO: East European Monographs, 1986).

Wyman, Mark, *Round-Trip to America: The Immigrants' Return to Europe, 1880–1930* (Ithaca, NY: Cornell University Press, 1993).

Ziegler, Béatrice, *Schweizer statt Sklaven: schweizerische Auswanderer in den Kaffee-Plantagen von São Paulo, 1852–1866* (Stuttgart: Franz Steiner, 1985).

Index

adoption: 131, 132
agents: *see* 'emigration agents'
agricultural labourers, English: alcohol, 94–95; allotments, 93–94, 134; diet, 93–95; isolation, 99; labour unrest, 91, 96, 99, 197–98; living conditions, 92, 93, 96; poverty, 92; regional differences, 92; working conditions, 92, 93; *see also* National Agricultural Labourers' Union
agricultural labourers' unions: *see* National Agricultural Labourers' Union
agriculture: family farms in Brazil, 15; land colonization in Latin America, 14–15, sharecropping, 28, 119; *see also* 'state colonies'
Aguiar, Luis H. F.: 41
alcohol: 19, 72, 74, 75, 79, 82, 86, 94–95, 132, 148–49
Algeria: 36, 153; *see also* 'French immigrants'
Almeida Portugal, Joaquim Maria de: 52–53, 54, 56, 57, 64, 67, 76
Alsop, Thomas: 105–6, 108, 109, 115, 116, 117–26, 194–95
Amazon: 40, 146
Anglo-Brazilian Times, The: 23–24, 25–27, 132–33; *see also* 'Scully, William'
Antonina (Paraná): 130
Apperley, J.W.: 114
Arch, Joseph: 89, 97, 99, 100, 196
Argelina (Paraná): 153
Argentina: 14, 18, 27, 37–38, 77, 87, 175, 200, 204, 278

Asia as a source of immigrants: 17; *see also* 'Chinese immigrants'
Ásmundsson, Einar: 20
Associação Central de Colonisação: 19
Assunguy (Cêrro Azul, Paraná): agriculture, 112–13, 157–62, 208, 212; administration, 156–57, 165, 172; buildings, 156–57, 158; churches, 168–69; class, 166–67; communications, 122, 130, 154–56, 155–62, 201–2, 208, 210–12; debts, 163–65, 168, 170; consular appeals, 171–72; employment, 130, 163–65; English and Irish in, 112, 114, 151, 157–70, 209; French in, 152–53, 157, 161, 165, 209; geography, 151–51, 202; Germans in, 152, 157, 165, 209, 209; health, 113, 168, 169–70; Italians in, 165, 209; market, 122, 162, 208; population, 165; schools, 169; Swiss in, 152, 209; *see also* 'Bariguy'
Australia: 100; *see also* 'Queensland'
Austrian immigrants: 175
Azores: 13, 15, 67
Baden: 20, 69
Bahia: 13, 14
Banbury: 119, 123, 195
Banbury Guardian, The: 123
Bariguy Lodge: 130, 133, 134
Barony of Forth (Wexford): 18
Barra do Pirahy (Rio de Janeiro): 128–29
Barrata, Senhor: 146
Barton, John:149
Bayliss, Thomas and Louisa: 111, 126

Belgian immigrants: 68
Beverley: 166
Bigg-Wither, Thomas P.: 150–51, 153, 155, 198
Birkenhead: 58–59
Birmingham: 56, 59, 67, 113, 114, 144, 215, 225
Black Country: 45–46; *see also under* 'Wednesbury'
Blumenau (Santa Catarina): 68, 152
Bocaiúva, Quintino: 22–23, 41, 42, 43, 203, 300
Bond, George, and family: 127–28, 130–33, 134, 134, 289
Bordeaux: 36
Brazil: as an 'exotic' destination, 7, 194; 'backwardness', 16; British merchants in, 14; British trade, 13–14; climate, 29, 108, 112; economic advantages, 25, 28, 108, 122; emigration terms, 53, 57, 103, 104, 106, 108, 109–10, 137–38; German image, 19; natural abundance, 28, 29, 30, 52, 108, 109, 112, 114, 115, 116, 122, 161, 166, 193; numbers of British immigrants, 10, 13; politics, 78; religion, 60; savageness, 30
Brazilian consulates: distribution of emigration literature, 28; Liverpool, 57, 60, 103, 105, 116, 199; London, 57; New York City, 41
Brazilian Emigration Agency (New York City): 41, 42, 43
Bremen: 68, 269
Bristol: 92
Britain, modernization: trade with Brazil, 13–14
British character: 25–26
British Guiana: 17
British West Indies: 17
Brown, William: 123
Brusque (Santa Catarina): 69, 207–8; *see also* 'Príncipe Dom Pedro'
Buckle, Thomas Henry: 29, 52
Buhlaw, Ernest: 139, 140, 142
Bushby, Elliot: 85–86
cachaça: 82, 86, 145, 148–49; *see also* 'alcohol'

Canada: 100, 101, 104, 107, 170, 172, 196, 213, 296
Cananéia (São Paulo): administration, 145–47; agriculture, 113–14, 121, 141–43, 213–14, 216; alcohol, 148–49; beauty, 112; communications, 139, 140. 143, 202, 213–14, 215, 216; diet, 123, 126; employment, 141, 146–47; health, 123, 126; immigrants from New York, 44; immigrants from England, 112, 114, 118–19, 120–21, 125–26, 144–45; land, 139–41; market, 143; population, 143–45, 215; produce, 114, 114, 115, 121, 141–42; *quilombo*, 215; school, 226; social life, 147–48; town, 139; *see also* 're-migration' *and* 'return migration'
Canary Islands: 13, 204
Casa de Saúde (Rio de Janeiro): 22, 43–44, 63–64, 65, 117, 118, 120, 125, 134, 174
Caseiros (Rio Grande do Sul): 44
Castle Garden (New York City): 22
Castro (Paraná): 202
Catholic Church: in Brazil, 60, 116, 144, 168; in Ireland, 35; in England, 46, 47, 48–49
Caymari, Bernardo: 41, 42
Cenci, Carlos Brent: 145, 146, 148, 149, 172, 200
Cêrro Azul (Paraná): *see* 'Assunguy'
Chamberlain family: 210
Cheltenham: 131
Chinese immigrants: 15–16, 17, 300
Cirencester: 110, 127, 131–32
Civil War, American: 39, 40
Clare, Angel: 3–5
Codman, John: 139
coffee plantations and immigrants: 16, 119
Commercial Agency of Brazil (London): 52–53, 54, 57; *see also* 'Almeida Portugal, Joaquim Maria de'
Confederates: 39, 40, 41, 42, 43, 139, 144
Connolly, John: 59, 63
Conservative Party (Brazilian): 78

Index

Cooke, Arthur: 106
Copenhagen: 21
Cotter, William: 17–18
Cottle, "Dr" Barziller: 70–72, 75, 76, 77
credit: 78, 163–65
cricket: 147
Crossley, John: 112–13
Cuba: 17
Curitiba: 69, 121, 130, 132–33, 153–54, 162, 164; *see also* 'Bariguy Lodge'
Davies, Bridget and James: 144, 149, 213–14, 215, 225–26
Davies, David and João: 215
Davis, John Frederick: 131, 134
debts: *see* 'credit'
demonstrations by immigrants: 80, 133
Desterro (Santa Catarina): 53, 75, 80, 85–86
Devon: 99
Dezenove de Dezembro, O: 133, 201
Dona Francisca (Santa Catarina): 68–69, 127, 152
Donovan, Father T.: 18
Dorchester: 122
Dorset: 97, 122, 166
Dundas, Charles: 199
Dunkirk: 269
Dunlop, Charles: 28–29, 52
Dutton, Thomas: 128
East Indians: 17
education: *see* 'schools'
Emery, Mary: 112
Emigrant Dept (Birkenhead): 58–59
Emigrants Hotel (Rio de Janeiro); *see* 'Casa de Saúde'
emigration agents: in Iceland, 20; in England, 10, 28, 52–53, 54, 76, 91, 103, 105, 109, 134, 137, 161, 284–85; in New York City, 41, 42, 43; *see also* 'Almeida Portugal, Joaquim Maria de', 'Alsop, Thomas', 'Haynes, Edward', 'Meadows and Christopher' *and* 'Yeats, William Ebenezer'
England: internal migration, 99, 100; Irish in, 44; rural conditions, 91–101; villages, 95–96

erva maté: 139
Espírito Santo: 119, 128–29, 286–87
Estcourt, James: 111
Eureka (Nevada): 77
experimental mobility: 9; *see also* 're-migration'
Fahy, Anthony: 37
Fairburn, Dr William: 175
Faro, Commendador: 128
Fell, Thomas: 108
Figueiredo, Dr F.C.: 146, 147
Fitz[gerald] family: 210
Fletcher, Daniel: 29
Florence Chipman (ship): 56, 58, 59, 60, 63, 66, 144, 225
Florianópolis: *see* 'Desterro'
Foreign Office (British): 81, 132
freedom: 101, 113
French immigrants: 19, 25, 151, 152–3, 157, 161, 164, 175, 273
Freyre, Gilberto: 4–5
Fulcher, Elijah: 149
Gentil, Carlos Perret: 15
Galvão, Ignâcio da Cunha: 74
Galway: 44
German immigrants: 16, 25, 59, 68–69, 118, 144, 145, 151, 152, 153, 157, 164, 174–75
German character: 19, 25–26
Germaness: 273
Germany: 19, 20
Gibbs, Joseph: 146, 149
Gloucester: 110, 111
Gloucestershire: 92, 97–98, 110–11, 105, 126, 127–28, 166; *see also* 'Cirencester'
Great Bourton: 123
Greenland: 20
Grenfell, John Pascoe: 18, 60
guide books: 28, 43, 52
Haher, John: 78
Hampton Chapel: 197
Harbury: 106
Harmer, George: 132
Haynes, Edward: 105, 117
Hardy, Thomas: 3–5, 93–94
Haskell, John: 84
Hawgood, Samuel: 122
Hayes, Peter: 83

Heinrick, Hugh: 48, 51
Hemming, John: 67
Hobsbawm, Eric: 91–92
Honduras, British: 40
Hopkins family: 84
Hunt, George Lennon: 23–24, 174
Hurst, James: 172–73
Hutchings, James: 131, 134, 135
Icelandic emigration to Brazil: 20–21
Iguape (São Paulo): 139, 144
Ilhota (Santa Catarina): 68
Illustrated London News, The: 98–99
Ireland: 35–36, 45
Irish: alcohol consumption, 19, 72, 75; character, 17, 19, 26–27, 72; habits, 51
Irish-Americans in Santa Catarina: 72–75
Irish in Algeria: 36
Irish in Brazil: backgrounds, 14; Brazilian opinion, 64; mercenaries, 17–18; *see also* 'Monte Bonito' *and* 'Príncipe Dom Pedro'
Irish in England: 44–58
Irish in Mexican Texas: 37
Irish in New York City: 39–40
Irish in Rio Grande do Sul: 18–19
Irish in Spain: 36
Irish in Wednesbury: *see* 'Wednesbury'
Irish language: 45, 48, 60
Itajaí (Santa Catarina): 69, 72, 77, 86
Italians immigrants: 21, 165
Jackson: Joseph: 149
Johnson, William: 149
Joinville (Santa Catarina): *see* 'Dona Francisca'
Kaisner, Augusto: 164
Kidder, James: 29
Kilchen, Joseph: 144
Kirby family: 84
Kitto, Woodward and Clarke: 56, 58
Klitzing, Frederick von: 79, 80
Kolonie Zeitung, Die: 74
Labourers' Union Chronicle, The: 100, 103, 104, 116, 121, 122, 123, 196
Lagoa Vermelha (Rio Grande do Sul): 44
Laguna (Santa Catarina): 67
Lancashire: 92

land: *see* 'agriculture' *and* 'terras devolutas'
Lander, Jane: 131
Lane, Albert: 111
language: English, 79, 193; Irish, 45, 48, 60; Portuguese, 29, 76, 119, 194
Latin America, colonization schemes in: 14–15
Lazenby, Joseph: 53, 54, 75–76, 80
Leamington Chronicle: 112
Leamington Spa: 97
Leão, Joaquim Antão: 78, 200, 298
Leopoldina (Bahia): 16
letters, emigrants': 112, 113, 115–16, 122, 123, 193, 194
Liberal Party (Brazilian): 78
library, lending: 147–48
Limerick: 23
Lincolnshire: 159
Lisbon: 118
Liverpool: 27, 117, 128; *see also* 'Birkenhead' *and* 'Brazilian consulates'
London, emigration to Brazil: 56, 137, 138, 166, 167, 169
Louisiana: 100, 197
Lusitania (ship): 118, 120
Luxembourg, emigration from: 269–70
Lyons, Edward: 113, 114, 115, 116, 144, 147
McCornville, Michael: 148
McDermott, Thomas: 84
MacMahon, Edmé Patrice: 36
Madeira: 13, 15
Manitoba: 296
Mansfield, Charles: 29
marriage laws, Brazilian: 24
Marseilles: 36, 153
Mathew, George Buckley: 81, 91, 111, 147, 173, 199
Maximillian, Emperor of Mexico: 79
Mayo: 44
Meadows and Christopher: 54, 56, 76, 112, 137–38, 166
Mello, Elpídio de: 77, 79, 119
Mendes (Rio de Janeiro): 126
Mendoça Franco, Melchior Carneiro de: 105
mercenaries: 17–18, 59, 68

merchants, British: 14
Merry, Alfred: 113
Meston, Manzilla: 105, 106, 108
Mexico: 37, 40
Middleton Cheney (Northamptonshire): 113
Miller, Kerby: 35
Minas Gerais: 16
Ministry of Agriculture (Brazilian): 65, 69, 74, 78, 83, 87, 118, 126, 162, 172, 200, 202
Ministry of Foreign Relations (Brazilian): 200–1
Minnesota: 100
Mitchell, John: 11, 213
Monks, Richard: 149
Monte Bonito (Rio Grande do Sul): 18–19
Montebello, Fazenda, (Espírito Santo): 119, 128–29, 286–87
Montgomery, George: background of, 46; finances, 46, 277; health of, 87; 'New Irelands', 61; on Brazil, 51–52, 54; on Catholics in England, 49–50, 54; on education, 49, 50; on Irish exile in England, 49; on the United States, 50, 51; Oregon emigration scheme, 50; petition to Pope Pius IX, 219–224
Moog, Víanna: 16
Moya, José: 204
Murphy, Cornelius: 149
Murray, Michael: 149
Napton-on-the Hill: 105–8, 116, 195–96
Nation, The: 49
National Agricultural Labourers' Union (NALU): 97–101, 105, 106–7, 128, 196–97; *see also* 'Arch, Joseph', *'Labourers' Union Chronicle*' and 'Yeats, William Ebenezer'
Nevada: 77
'New Ireland': 60, 78, 87
New York: 39–41, 50, 192
New Zealand: 100, 104, 107, 172
Nichols, John: 149
North America (steamship): 43
Northamptonshire: 113
Nova Câmbria (Rio Grande do Sul): 18

Nova Friburgo (Rio de Janeiro): 16
Nova Petrópolis (Rio Grande do Sul): 44
Ohio: 50
Official Agency of Colonization, Brazilian: *see* 'Ministry of Agriculture'
Ontario: 101
onward migration: *see* 're-migration'
Oregon: 50
Orgill, Austin: 85
Orgill, Martha Connolly: 84–85
Oscott College: 46
Québec: 170
Oxfordshire: 105, 114, 117, 119, 122, 123–24, 195
Pailton: 112
Pará: 5, 14
Paraguay: 300
Paraguayan War: 59, 78
Paraná: climate, 28; communications, 150–51, 153–4, 201, 202; immigrant settlement, 151–56; migrants from Santa Catarina, 69, 87, 152; population, 151; riches, 109, 201; Scully's opinion of, 66; social structure of, 149–50; trade, 150
Paranaguá: 121, 130, 133, 135, 153
Parliamentary questions, British: 200
Passenger Act, 1855: 10
Patagonia: 14, 18, 278
Patagonia (ship): 128
Pedro I, Dom, emperor of Brazil: 18
Pedro II, Dom, emperor of Brazil: 51–52, 63–64, 122, 123
Pelotas (Rio Grande do Sul): 18
Pennsylvania: 50, 85
Pernambuco: 13, 14
Phipps, Constantine: 120, 125
Piddletown: 122
Piúma (Espírito Santo): 119
Pius IX, Pope: 219–24
Polish immigrants: 151, 175
post: 161; *see also* 'emigrants' letters'
Pius IX, Pope: 47, 49
Porto Alegre (Rio Grande do Sul): 44
Portugal: 13, 19
Portuguese court, transfer to Brazil: 13
Portuguese immigrants: 15, 21, 25

Portuguese language: 29, 76, 119, 194
Príncipe Dom Pedro, (Santa Catarina):
administration, 70–72; agriculture,
69, 76, 82; Catholic Church in, 53,
75; communications, 69, 71, 76;
corruption, 71–72, 77, 79–80, 82;
credit, 78; dissolution, 207–8;
disorder, 72, 74, 75, 79, 80, 202;
education, 76; employment, 72, 77,
79, 80; English in, 76–77;
establishment, 69; ethnic tensions,
79–80; evacuation of, 85–85;
expulsion of immigrants, 72, 74, 79;
floods, 69, 82, 83–84, 85; Germans
in, 72; hunger, 78; immigrant
arrivals, 73–74; Irish in, 44, 53,
72–87; land allocation, 74, 80;
markets, 82; physical description, 69;
re-migration from, 77, 81, 85, 86–87,
166; temperance movement, 75;
timber, 74; urbanization, 76, 83;
Xokleng, 74
Preston, George: 174
Protestantism: 22, 24, 27, 31, 34, 36,
39, 41, 50, 59, 60, 61, 63, 116, 148,
168–69, 216
Prussia: 20; *see also* 'Germany'
Québec: 101
Queensland: 109, 112
quilombo: 215
racial determinism: 30
railways: 86
Rainbow, William: 106
Randall, James: 173
Randall, Mrs: 170
Recife: 13
Redman, Joseph: 168–69
Refugio (Texas): 37
re-migration: 9, 77, 85, 86–87, 127,
152, 166, 172, 270, 276–77
Renaudin, Joseph: 170
return migration: 9, 8, 25, 82, 126–27,
135, 173–75, 273
Rev. G. Montgomery's Register, The:
46–47, 219
Rifleman (ship): 213–14
Rio de Janeiro: British business, 14;
destitute Britons and Irish in, 81,
84–85, 126–27, 172–73, 175, 199;
English immigrants' enjoyment of,
119–20; German immigrants, 118;
health, 126; immigrant port of entry,
21; labour market, 23, 65–66;
Wednesbury immigrants in, 65,
84–85; *see also* 'Casa de Saúde'
Rio Grande: 14
Rio Grande do Sul: 44; cattle ranching,
14, Irish in, 18–19; Welsh in, 19
Robinson, William: 172
*Royal Leamington Spa Courier
and Warwickshire Standard, The*:
195, 196
Rudé, George: 91–92
Ryan, William: 159
St Patrice (Algeria): 36
Sallentein, Franz: 74
Salvador (Bahia): 13
San Patricio (Texas): 37
Santa Rosa (ship): 125, 126, 127
Santa Catarina: 43, 67, 118;
agricultural colonies in, 68–69;
agriculture in, 66; climate of, 28;
economy, 68; Indian resistance, 67,
74; land, 68; migration to Paraná, 69,
87, 152; population concentration, 67
Santarém (Amazonas): 40
Santos (São Paulo), destitute Britons
in, 81, 85, 86, 87
São Francisco do Sul (Santa Catarina):
67
São Jerônimo (Rio Grande do Sul): 18
São Paulo: 40, 87, 109
São Pedro de Alcântara (Santa
Catarina): 68–69, 72
Sardinia: 19
Schleswig-Holstein: 69
Schneeburg, Maximilian von: 72
schools: 76, 148, 169, 226
Scully, William: 22, 23, 58, 64, 66,
119, 175, 192, 202–3, 272–73; *see
also 'Anglo-Brazilian Times' and
'guide books to Brazil'*
Sheasby, Thomas: 195
Shotteswell: 123
Somerset: 111
sharecropping: 28, 119, 128–29
Sheasby, Thomas: 106
shipping: 53, 104

slavery and slaves: 14, 15, 16, 40, 67, 78, 194, 198; *see also* 'white slavery'
Sociedade Internacional da Imigração: 21–22, 41; *see also* 'Casa de Saúde'
Somerset: 96, 173
Southey, Robert: 33
Spain: 13, 36
Staffordshire: *see under* 'Birmingham', 'Black Country' *and* 'Wednesbury'
Standard, The: 52
Stanton, Mary: 123–24
state colonies: 40–41, 44, 70, 108–9, 118, 122
Stroud: 98
Sturz, Johann Jacob: 64, 75
Swedish immigrants: 175
Swiss immigrants: 15, 16, 19, 144, 152
Tamplin, Charles and Caroline: 167–70, 294, 297
Tavares Bastos, Aurélio Cândido: 22–23
Taylor, Henry: 196–97
Taylor, Philip: 9
tea plantations: 16
terras devolutas: 68, 69, 141
Tess of the D'Urbervilles: 3–5, 93–94
Tetu, Ubald: 170
Texas: 37, 100, 144, 146
Tigar, Frederick Tertius: 89, 161, 164, 165, 166–67, 296, 297
Times, The: 76, 195–96. 197–98
Trinidad: 17
United States: 17, 19, 20, 23, 26–27, 35, 40, 42, 100, 193; *see also* 'New York', 're-migration' *and* 'return migration'
United States and Brazil Steamship Company: 41, 42
Universal News, The: 30, 33, 60, 87
Uruguay: 18, 77
Vanderhoff, Senhor: 146
Venezuela: 40
Virginia: 100
Votuverava (Paraná): 155–56, 211
Waidell, Dr: 83
Walker, Reuben: 111
Warwickshire: 97, 105–8, 118, 122, 166, 195, 197

Watson, Charles: 80
Watts, William: 146, 148
Wednesbury: Brazil, 54–56, 59–60, 63, 65, 67, 77–78; economic conditions, 47, 219–24; Irish in, 44, 45, 46, 48, 219–24; Irish language, 48; poverty in, 46, 48, 219–24; St Mary's Church, 46, 54; sectarianism in, 47–49, 219–24;
Wellesbourne: 106, 107
Welsh immigrants: 18, 118, 119, 278
Werneck, Luiz Peixoto de Lacerda: 17, 19
West Indies: 17
Wexford: 18
white slavery: 131–32, 199
'whitening' of the Brazilian population: 16–17
Wilts and Gloucestershire Standard, The: 131–32, 193–94
Wiltshire: 93
Wolf and Company, John S.: 56
women, employment prospects: 23
Würtemburg: 20
Xoxleng: 67, 74
Yates & Company: 128
Yeats, William Ebenezer: 97–98, 105, 109–11, 125, 127, 196
Young, Edward: 113
Yorkshire: 92, 166

www.ingramcontent.com/pod-product-compliance
Ingram Content Group UK Ltd.
Pitfield, Milton Keynes, MK11 3LW, UK
UKHW041228200426
11947UKWH00034B/396